Death Valley: GEOLOGY, ECOLOGY, ARCHAEOLOGY

Charles B. Hunt

Death Valley

GEOLOGY, ECOLOGY, ARCHAEOLOGY

UNIVERSITY OF CALIFORNIA PRESS
Berkeley   Los Angeles   London
1975

University of California Press
Berkeley and Los Angeles, California
University of California Press, Ltd.
London, England

Library of Congress Catalog Card Number: 74-84094 .

Designed by Harlean Richardson

Printed in the United States of America

FRONTISPIECE:

The Basin and Range Province, of which Death Valley is a part, consists
of broad, down-faulted valleys between block-faulted mountains. This view,
northeast from Telescope Peak, shows the east slope of the Panamint
Range in the foreground, and gravel fans built of debris eroded from the
mountains sloping to the floor of Death Valley. Beyond is the faulted front of
the Black Mountains (right) and Funeral Mountains (left). On the skyline is
the Amargosa Desert and block-faulted mountains in and around it. (Photo-
graph by Warren B. Hamilton.)

Affectionately dedicated to
Matt and Rosemary Ryan,
formerly of the Park Service,
known to their many friends as
Mr. and Mrs. Death Valley

Contents

1 What Is Death Valley?

Death Valley has an emotional impact on almost everyone who visits it. To some the valley is a place partway to hell, a hot, eerie salt flat below sea level, partly enclosed by mountains that rise as high as 2 miles above the salt. To others it is a place for desert vistas across mountaintops—to the Spring Mountains more than 50 miles to the east and almost 12,000 feet high and to the Sierra Nevada more than 50 miles west with peaks above 14,000 feet. To still others it is a land of ever-changing colors from sunrise to sunset, or a land of desert flowers and animals, or a land of fascinating history and archaeology. And most visitors wonder what Death Valley is, and how it became what it is (fig. 1).

THE LANDSCAPE

Death Valley is just one of the many desert basins between mountain ranges (fig. 2) in the western region known as the Basin and Range Province, an area that includes most of Nevada, western Utah, southeastern California, southern Arizona, south-western New Mexico, and trans-Pecos Texas. Each of the basins

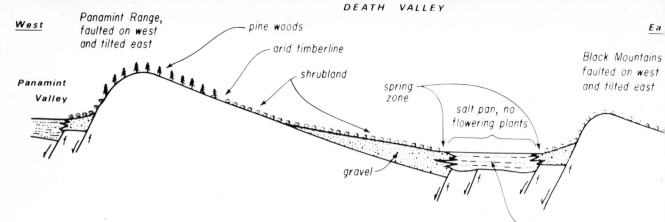

West

Panamint Range, faulted on west and tilted east

pine woods

arid timberline

shrubland

spring zone

salt pan, no flowering plants

Ea

Black Mountains faulted on west and tilted east

Panamint Valley

gravel

beds of silt and salt

FIG. 1. Simplified cross section of Death Valley. The valley is a structural sag between two mountain blocks, Panamint Range on west and Black Mountains on east. Faults (*f*) along steep west fronts of mountains have lowered the valleys with respect to the mountains. The downfaulted valleys collect sediment eroded from mountains; coarse gravels form fans sloping to floor of valley and silt and salt are deposited on playa. Groundwater is shallow where gravels grade into playa beds, and a zone of springs sustains growth of water-loving plants. The playa is too salty for flowering plants. Desert shrubs grow on fans and up to about 6,000 feet on mountain slopes. Above this arid timberline, pine trees grow.

and ranges, including Death Valley, is characterized by three very different environments.

Bordering Death Valley are high, rocky mountain ranges; beyond them lie other desert basins. Sloping to the valley floors from the base of each range are gravel fans built of debris washed from the mountainsides. The gravel fans end at the edge of a broad, salt-crusted mud flat which is the dry bed of a Pleistocene lake, referred to as a playa. The mountains, the fans, and the playa, though closely related, are not at all alike.

Death Valley is young and its principal topographic features reflect the recency of the earth movements that uplifted the bordering mountains and caused the valley to sag between them. In fact, the earth movements that produced the valley and mountains are continuing, a subject discussed in chapter 6. In mountains that are geologically old, like the Appalachians, valleys are the result of erosion. Not so Death Valley nor the valleys along the Amargosa River; they are the result of earth movements—folding and faulting of the rock formations.

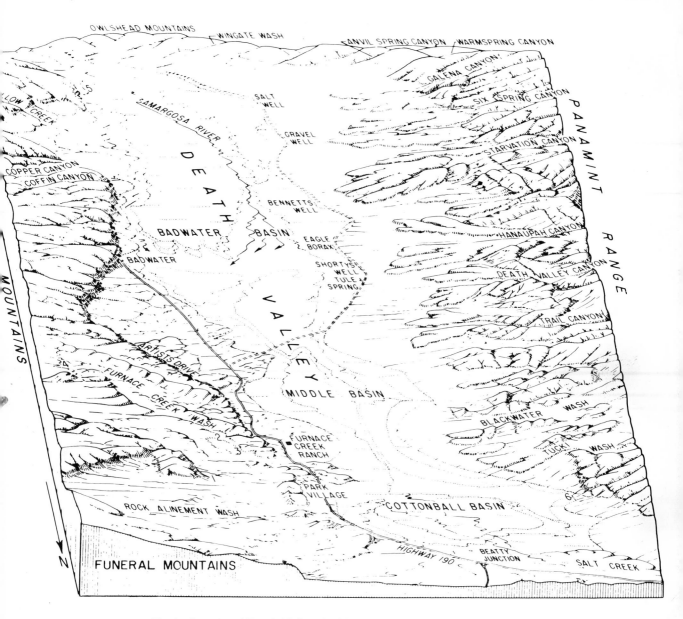

Block diagram of Death Valley, looking south. 1. Nevares Spring. 2. Travertine Spring. 3. Texas Spring. 4. Corkscrew Canyon. 5. Coyote Hole. 6. West Side Borax Camp (Shoveltown). (From U.S. Geol. Survey Prof. Paper 494-A.)

Preface

From 1955 to 1960 I had the opportunity to study the geology of Death Valley for the United States Geological Survey. The results of my work are presented in three of the survey's professional papers (all published in 1966 by the United States Government Printing Office in Washington): 494-A (with Don R. Mabey), *Stratigraphy and Structure, Death Valley, California;* 494-B (with T. W. Robinson, Walter A. Bowles, and A. L. Washburn), *Hydrologic Basin, Death Valley, California;* and 509 (with L. W. Durrell), *Plant Ecology of Death Valley, California.*

At the same time, my wife, Alice Hunt, made an archaeological survey for the Park Service under the auspices of the Department of Anthropology, University of Southern California, Los Angeles. Her study was published in 1960 by the University of Utah as no. 47, Anthropological Papers, under the title *Archeology of the Death Valley Salt Pan, California.*

The three professional papers of the Geological Survey are out of print, and there is no likelihood of their being reprinted because of the expensive colored geologic maps. This book is intended partly to fill the gap thus created and partly to reach a broader audience by presenting the essence of all four reports in as nontechnical language as possible.   C. B. H.

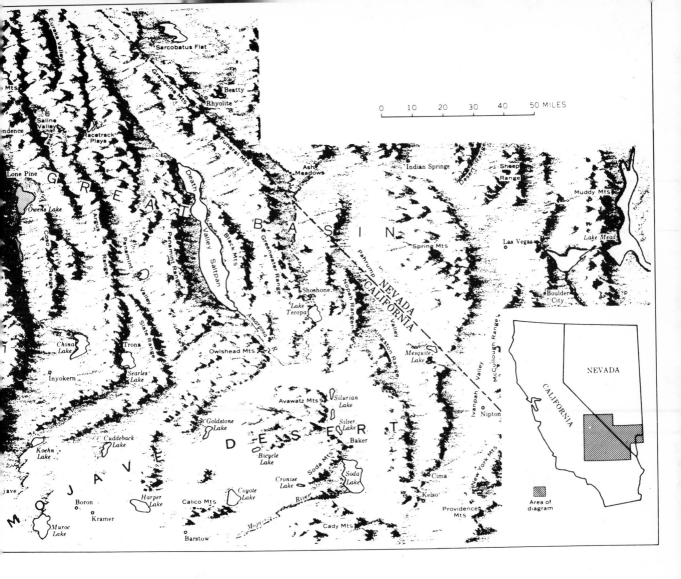

FIG. 2. Death Valley, in southern part of Great Basin and just north of
Mojave Desert, is midway between Sierra Nevada on the west and Colorado
Plateau, which lies just east of Lake Mead. (From U.S. Geol. Survey Prof.
Paper 494-A.)

SALT

The Death Valley playa, covering more than 200 square miles, contains one of the world's largest salt pans (chap. 3). The salts are of various kinds, but sodium chloride is most abundant. It is like table salt but less refined, some of it downright muddy. The salt crust, ranging from a few inches to a few feet in thickness, rests on damp mud. There are no flowering plants and almost no animal life; all the water on the salt pan is saltier than seawater and is quite undrinkable. How much salt is contained in the salt pan and buried in the several thousands of feet of mud beneath it is not known, but it is more than enough to make a pile a mile in diameter and as high as Telescope Peak (11,000 ft.).

Had the prehistoric Indians realized the worth of their salt resource to the rest of the world, their lot might have been different. Salt served as money in ancient and medieval times and was a major source of revenue, but the profit was said to be "more to the king than to the makers and sellers." Although salt was a state monopoly, the Death Valley Indians could have broken that market, but perhaps they needed a trucking company or an airline.

Salt was so scarce during colonial days that it was exempted from the Navigation Acts of 1663, and the colonies could import it directly from Spain or Portugal. As late as 1800, debts could be paid legally with a compound unit that was mostly produce, "half in meat, whether beef, pork, bear, or venison; one-fourth of corn; one-eighth salt; and one-eighth money." Death Valley salt, unlike gold, was in safe storage and needed no security guards.

WATER

Drinking water in Death Valley is available from three sources. In the mountains are small springs, mostly just seeps. In a sandy

zone bordering the salt pan, the groundwater is close enough to
the surface to be reached by hand-dug wells; most of this water
is too salty for drinking, but on the west side, under Telescope
Peak, where the recharge of fresh water to the valley is maximum,
the groundwater is fresh, as at Tule Spring, Shorty's Well, Eagle
Borax, and Bennetts Well. Most of the drinking water in Death
Valley is supplied by large warm springs discharging along rock
fractures, or faults. As explained in chapter 2, the source of the
water in the springs in the Furnace Creek area is far to the east,
in the Spring Mountains. The springs are large enough to irrigate
a plantation of date palms and a golf course at Furnace Creek.

SAND DUNES

Sand dunes are prominent features of many desert landscapes. In
Death Valley dunes are extensive north of the salt pan, along Salt
Creek and at Mesquite Flat, and south of the salt pan along the
Amargosa River. In the driest part of the valley, around the edge
of the salt pan, dunes are infrequent except near the springs on
the west side opposite Telescope Peak which discharge fresh
water. The rest of the ground is securely cemented with salt
(fig. 3).

GRAVEL FANS

The gravel fans are discussed in detail in chapter 4. Fans on the
west side of the valley are huge, some 6 or more miles long and
1,500 feet higher than the salt pan. Fans on the east side are
comparatively small, less than a mile long and apexing only a
few hundred feet above the level of the salt pan. This difference
reflects variations in the structural setting. Other dissimilarities

between fans reflect differences in the kinds of rocks that were eroded to form them or differences in geologic age. Gravel fans are very dry, in Death Valley and elsewhere in the Basin and Range Province, for the ground is permeable and water is quickly lost by seepage. Even the fans built of mud are dry because they are nearly impermeable. All these differences are closely reflected in the plant and animal geography.

MOUNTAINS

The mountains bordering Death Valley are young; in fact the earth movements that formed them are still continuing (chap. 6). There are three different kinds of mountainous terrain, depending on the type of rocks in the mountain. At the south end of the Panamint Range are Precambrian rocks that erode to form smooth hillsides even though slopes are steep. Farther north in the range and in the southern part of the Funeral Mountains the rocks are mostly Paleozoic limestone formations marked by exceedingly rough cliffs, jagged peaks, narrow canyons, and sharply angular rock fragments. The roughest and least hospitable mountains are

FIG. 3. Despite the aridity, there is little dune sand on Death Valley fans because most of the ground is cemented with salt. At this hill, opposite entrance to Artists Drive, sand has drifted northward. View is south.

formed of Tertiary volcanic rocks and playa deposits, like those at the north end of the Black Mountains. These colorful rocks are intricately gullied and eroded into gigantic badlands, far bleaker than the scenes in Badlands National Monument.

The Tertiary formations are nearly impermeable, and rainwater runs off quickly. The ground is practically without vegetation and is avoided by burros and mountain sheep. The limestone terrain is favored by sheep; burros prefer Precambrian geology, since limestone would cut their hooves.

CLIMATE

Death Valley is the hottest and driest part of the Southwestern desert. Winter temperatures on the valley floor rarely dip to freezing; summer temperatures average more than 100°F and have reached a maximum of 134°.

Annual precipitation averages only about 1.5 inches, and twice in the half century of weather records there has been no recorded rainfall during the twelve months. Twice the annual rainfall has exceeded 4.5 inches (fig. 8). Fluctuations in precipitation can be extreme, from 3.4 inches in 1953 to zero in 1954. The record may be divided into ten-year periods with substantially different averages: 2.63 inches in 1936-1946 but only 0.83 inches in 1924-1935. Over longer periods the fluctuations have been much wider; during the two or three millennia preceding the Christian era, Death Valley contained a lake 30 feet deep. In the Pleistocene there was a lake 600 feet deep.

Precipitation is heaviest in winter and lightest in summer, as is true of most of California. Above 5,000 feet it is several times greater than on the valley floor. During the biblical period when it rained for forty days and forty nights, Death Valley is said to have had a quarter inch.

The aridity is magnified by the high rate of evaporation which, on the valley floor is a hundred times precipitation. Water losses from saline ground, however, are surprisingly slow; brines and wet saline muds dry much more slowly than does fresh water. The loss from 200 square miles of mud flat is less than one might imagine.

Ground surface temperatures are very much higher than air temperatures; a maximum of 190°F has been recorded. Yet even where surface temperatures are highest, the ground is much cooler a few inches below the surface. At a depth of 1 foot the temperature varies only slightly from the seasonal average and at a depth of 4 feet, only slightly from the average annual air temperature. These temperature conditions within the ground are important factors controlling plant and animal life. In a cold climate cellars protect against freezing; in a warm climate they protect against summer heat.

During still nights, air temperatures on the floor of Death Valley may be 10°F cooler than on the fans a thousand feet higher. This phenomenon is not due primarily to cold air draining into the valley, because even at the canyon mouths temperatures are higher than on the valley floor. The cooling is attributed to evaporation from the salt pan, mostly around the edge, and to transpiration of plants growing there. The layer of cooled air may be as thick as 300 feet.

Unlike the scanty and uncertain rainfall, winds in Death Valley are strong and dependable. The winds are very likely to force campers out of the campgrounds because of violently blowing sand and silt.

J. Ross Browne, author of the first United States mineral resource report (1868), wrote about the region: "The climate in winter is finer than that of Italy . . . [though] . . . perhaps fastidious people might object to the temperature in summer. . . . I have even heard complaint that the thermometer failed to show the true heat because the mercury dried up. Everything dries; wagons dry; men dry; chickens dry; there is no juice left in anything, living or dead, by the close of summer."

PLANT AND ANIMAL GEOGRAPHY

Death Valley, a land of extremes, is well suited for the study of plant and animal geography because of the wide range in climate and type of ground. Temperature controls the length of the growing season, as shown by the altitudinal zoning of plants and animals on the mountains. The effect of extreme temperatures is evidenced by the barren ground at the surface, which is subject to very high temperatures.

Different species of plants have different capabilities for resisting drought, and their distribution brings out striking variations in the kind of ground and its water-holding capacity. Plants requiring shallow groundwater where their roots have a perennial water supply have different salinity tolerances and respond to variations in the quality of the water.

There are similar but less evident differences in animal geography. Reference has already been made to differences in the kind of ground preferred by burros and bighorn sheep. Most of the pools of water evaporate too quickly or are too salty for fish, but some fish live in pools on the west side of Cottonball Marsh. Rabbits, rodents, and insects are plentiful in the zone of shallow groundwater bordering the salt pan; they are much less numerous on gravel fans and are virtually nonexistent on the still drier and bare badland hills.

PEOPLE

There is no certainty as to who first visited Death Valley, or when. The earliest immigrants of record were hunters, who came perhaps 10,000 years ago. Judging by their tools, these early visitors hunted big game. Later, about 2,000 years ago, Indians lived by the shore of the 30-foot lake. They hunted smaller game and gathered seeds. For a discussion of these people, see chapter 8.

The first white people known to have visited Death Valley arrived in 1849, in an episode related to the California gold rush (see chap. 9). Early in the 1880s borax was discovered in Death Valley, giving rise to the well-advertised 20-mule teams that hauled the product to the railroad at Mojave. Mining and other aspects of Death Valley history are summarized in chapter 7.

Present residents of Death Valley include members of the Park Service who manage the national monument, persons engaged in providing facilities and services to visitors, and those concerned with mining properties, notably the lands containing talc or borax. There is a Death Valley, California, post office, and a Death Valley elementary school, but children in the higher grades travel by bus to Shoshone, 50 miles away. The 100-mile round trip must be one of the longer school bus routes in the country.

There are two museums, one operated by the Park Service and one by the Fred Harvey Company. At either Stovepipe Wells or Furnace Creek visitors may camp, rent comfortable cabins, or obtain luxury quarters at the inns. At these centers there are stores that sell groceries and gasoline.

2 Water

Few people other than hydrologists take water seriously; most people take it for granted. Even in Death Valley, dry as it is, the turn of a faucet at an inn or motel produces water, when and as it is wanted, hot or cold. Water comes easily and inexpensively despite the aridity. Yet there were times, only a few thousand years ago, when Death Valley had such excesses of water it contained lakes. About 2,000 years ago the floor of Death Valley was flooded by a pond 30 feet deep. Several thousand years before that, during the Pleistocene, Death Valley had lakes hundreds of feet deep. The number and details of the episodes that produces the lakes are obscure, but the lakes left an indelible record of now dry shorelines at many places around the Valley.

LAKES, WET AND DRY

Pleistocene Lakes   During Pleistocene glaciations, when moisture was more abundant than it is today, the desert basins of southeastern California, Nevada, and western Utah contained lakes. Pleistocene Lake Bonneville, ancestor of the Great Salt Lake, was 1,000 feet deep and covered most of western Utah. Such lakes eroded shorelines and built gravel embankments where the water lapped aginst gravel fans. At the mouths of canyons deltas were formed, extending out into the lakes. Salt Lake City and Provo are both built on such deltas. There were probably several Pleistocene lakes in Death Valley, but the record is not sufficiently complete to reconstruct their histories or to know, in fact, how many there were. One of them, whose maximum depth was 600 feet, has been called Lake Manly, but it is better known as the Death Valley Pleistocene lake.

Only a few widely separated places on Death Valley gravel fans have distinct lake deposits. One easily accessible location readily identifiable as a lake bar is crossed by the road to Beatty, 2 miles north of highway 190 (fig. 4). It is 120 feet above sea level, that is, about 400 feet higher than the floor of Death Valley. One can stand on that bar, look south, and imagine a lake 400 feet deep in the valley. Other well-preserved lake remnants on the gravel ridge extending north from the Park Service residential area (fig. 5) can be seen plainly from highway 190, about a mile to the west. Some of the remnants are 200 feet above sea level, but they may have been deposited at a lower altitude and faulted up to their present position. As yet no one can be sure how much structural movement has occurred since the lake beds were deposited.

Another accessible beach deposit is on the spur just north of the exit from Artists Drive. It is below sea level. Stone artifacts on top of the gravel have been interpreted by some to record Indian occupation while the lake was there, but this interpretation has doubtful validity. Artifacts on the gravel, even if part of the desert pavement, could have been dropped there long after the lake was gone. No unequivocal artifacts have been found within the deposits, which would indicate contemporaneity. Moreover, the unequivocal artifacts that have been found are like those of Holocene Indians and are part of the desert pavement and

FIG. 4. Beach bar built by early Death Valley lake about 150 feet above sea level and 400 feet above valley floor. It is crossed by road to Beatty 2 miles north of Beatty Junction.

FIG. 5. Shoreline of old Death Valley lake, 200 feet above sea level and 450 feet above the valley floor, forms a horizontal terrace across hill in center of picture. It is on a ridge of faulted gravels a mile north of Park Service residential area.

therefore younger than the gravel deposits. Gravels deposited in beach bars are cleanly sorted and shingled and not at all like the poorly sorted, irregularly bedded gravels deposited by streams on the gravel fans.

Other strand lines, lime-cemented layers plastered against the front of the Black Mountains, may be seen as terraces at Mormon Point and Shoreline Butte in the southern part of Death Valley. The only trace of shoreline found along the west side fans is on the hill of basaltic lava at the end of the ridge dividing Tucki and Blackwater washes. The slight record left by early Death Valley lakes may mean that they were short-lived, with fluctuating lake levels. Whatever the cause, these lakes have left as slight a record as any in the Great Basin.

The Pleistocene Death Valley lakes have been attributed to overflow down Wingate Wash from a lake formed when Panamint Valley was flooded by overflow from Searles Lake and other lakes farther up the Owens River. But Wingate Wash contains no record of such flooding, and probably the water in the Death Valley Pleistocene lakes came from the south by way of the Mojave River at Soda Lake. The distribution of fishes in desert springs in the region suggests the same thing.

Holocene Playas and Lakes   As the Pleistocene waned, roughly 10,000 years ago, Death Valley became a dry lake bed, or playa. The valley floor must have been a mud flat, subject to seasonal flooding; probably it had no more salt crust than exists today on parts of the valley floor where seasonal flooding occurs. During the early Holocene the climate became drier and warmer than it is today, and the hydrologic regimen then, as now, must have been conditioned more by groundwater than by surface water. It is believed that a salt crust began forming during the dry period

During the middle Holocene the climate was wetter than it is today. There was alluviation along the Amargosa River, and the floor of Death Valley was flooded by a shallow lake, or pond, that reached a maximum depth of 30 feet (see fig. 6). Although the pond was probably of brief duration, it must have reworked the salt crusts that had formed during the early Holocene. This shallow lake has been dated archaeologically in the period from 3000 B.C. to A.D. 1 (see chap. 8). Desiccation of this middle Holocene lake gave rise to the salt pan; as the salt pan developed, the floor of Death Valley was tilted eastward with the result that the eastern shoreline of the lake is about 20 feet lower than the western shoreline (see chap. 6).

Death Valley's Most Recent Lake   The most recent lake in Death Valley was formed in February 1969, when a thaw in the Panamints, which melted the snow, was accompanied by scattered warm rains lasting two or three days. Salt Creek and the Amargosa River discharged all the way to Badwater Basin. About 80 square miles of the salt pan was flooded; in the lowest part, between Badwater and Eagle Borax, the water was 2 to 3 feet deep. If the average depth over the 80 square miles was 1 foot, the volume of water was about 50,000 acre-feet. This amount would represent runoff of about 0.1 inch from the whole hydrologic basin that drains to Death Valley. The 1969 lake provides a measure of what flood conditions must have been like to develop the far deeper lake in middle Holocene time and the still deeper Pleistocene lakes.

The Holocene lake was ten times as deep as the 1969 lake and covered at least four times as much surface; the volume of water therefore must have been about forty times as large. We cannot be sure about the depth or the extent of the Pleistocene lakes because of changes in the valley floor, but the volume of water in them must have been at least fifty or, more probably, a hundred times greater than in the Holocene lake.

Salty brines evaporate slowly, and although the level of the 1969 lake dropped rapidly, water stood on the floor of the valley for many months (fig. 7). The flooding eroded and redistributed large areas of the salt crust.

QUANTITY AND QUALITY OF DEATH VALLEY WATER

The water budget for the whole earth is a balanced economy: precipitation is equal to evaporation plus the water lost by transpiration of plants. The water budget for Death Valley is also a balanced economy, but the water there divides very differently from the average for the whole earth. Precipitation on land divides four ways:

1. For the whole earth, one part, amounting to perhaps half the total precipitation, returns directly to the atmosphere by evaporation. In the hydrologic basin draining to Death Valley, evaporation losses probably exceed 90 percent.

2. For the whole earth, a second part, amounting to perhaps a sixth of the total, is returned to the atmosphere by transpiration. In the Death Valley hydrologic basin such loss is probably almost 10 percent; although vegetation is sparse, loss of water by transpiration is considerable because vegetation is concentrated at water sources.

3. A third of the earth's water discharges into the ocean; Death Valley has no such loss.

4. Precipitation that enters the ground amounts to perhaps 1 or 2 percent for the whole earth but substantially less than that in the Death Valley hydrologic basin.

Volumes of water can be measured in terms of acre-feet. Stream discharge is commonly expressed in units of cubic feet of flow per second, referred to as second-feet. Since 1 cubic foot equals 7.48 gallons, 1 second-foot is about equal to 7.5 gallons per second, 450 gallons per minute, 2 acre-feet per day, and 725 acre-feet per year.

Whereas people take the occurrence or availability of water for granted, they usually are sensitive about water quality, which depends on dissolved solids. Water that has no dissolved solids—chemically pure water—tastes flat; most people like a little salt. Too much salt can be distasteful, and the wrong kind of salt, like magnesium sulfate (Epsom salts), is bitter and acts as a cathartic.

FIG. 6. Shoreline of Holocene lake which was largely responsible for salt pan is conspicuously marked by upper limit of salts. This illustration shows a cove north of West Side Borax Camp.

FIG. 7. Standing water ponded along Salt Creek opposite Furnace Creek fan in 1969 flood. Since the water is a nearly saturated brine which evaporates slowly, such ponds may persist for a considerable period of time, despite high rate of evaporation in Death Valley.

Water quality is expressed as parts per million, or percentages, of the dissolved solids, mostly salts. Differences in quality are shown by an easily remembered scale that decreases in geometric progression: (1) 330,000 parts per million (33 percent) is saturation; (2) 33,000 parts per million (3.3 percent) is seawater; (3) 3,300 parts per million (0.33 percent) is too salty for human consumption but may be used for livestock; (4) 330 parts per million is the standard for municipal water systems. The quantity of dissolved solids in average city water amounts to somewhat less than a level teaspoon in a gallon of water, but Death Valley water contains a larger amount of salt.

SOURCES

Water in Death Valley comes largely from rain and snow in other places. As the saying goes, "You can leave the windows open; it is not likely to rain." Annual precipitation at Furnace Creek Ranch has ranged from zero in 1929 to more than 4.5 inches in 1913 (fig. 8). Precipitation increases with altitude; at 5,000 feet in Death Valley it averages about 6 inches, and at 10,000 feet it is probably twice that. Figure 9 shows the probable distribution of rainfall in the Death Valley hydrologic basin. Fortunately, precipitation is not the only source of water that reaches the floor of Death Valley. Almost half the water comes via faults that provide underground drains for mountains and valleys outside the hydrologic basin.

A principal source seems to be the Spring Mountains, the third range east of Death Valley and 50 miles away (fig. 2). These mountains, reaching to 11,912 feet at Charleston Peak, are wetter than the Panamint Range because they are higher and are farther out from the rain shadow cast by the Sierra Nevada. Meltwater from snow on the Spring Mountains recharges groundwater in Pahrump Valley, which is being overpumped for irrigating fields

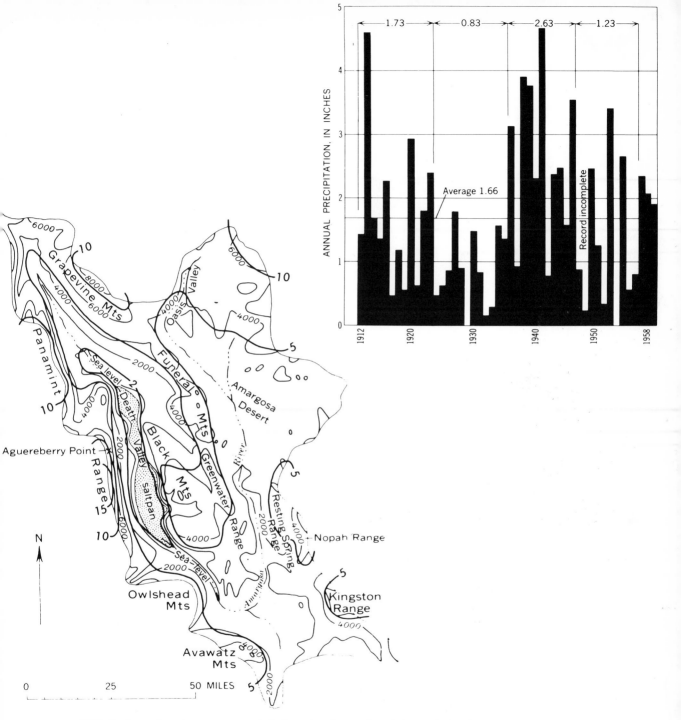

FIG. 8. Annual and average precipitation showing wet and dry periods at Furnace Creek Ranch. (From U.S. Geol. Survey Prof. Paper 494-B.)

FIG. 9. Sketch map showing probable distribution of rainfall in Death Valley hydrologic basin. (From U.S. Geol. Survey Prof. Paper 494-B.)

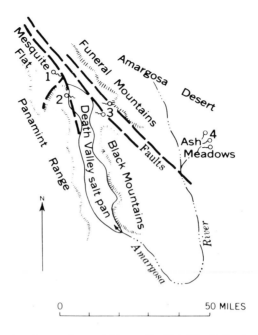

FIG. 10. Map showing springs at Mesquite Flat (1), Cottonball Marsh (2), Furnace Creek (3), and Ash Meadows in Amargosa Desert (4) and their relation to major faults in area. (From U.S. Geol. Survey Prof. Paper 494-B.)

of cotton, a surplus crop. Otherwise the valley would have an excess of water.

Pahrump Valley is a sieve, however; the water escapes from it via the many faults in the surrounding mountains. Some of this water reappears in huge springs in the Amargosa Desert and along the Amargosa River valley to the south. Faults along Furnace Creek extend from the Amargosa Desert to Death Valley, and groundwater moves from the Amargosa Desert to supply the big springs at the mouth of Furnace Creek (fig. 10). That this groundwater is the source of water in Death Valley springs is indicated by their chemistry (fig. 11) and by the fact that catchment areas at the springs are not sufficient to provide the

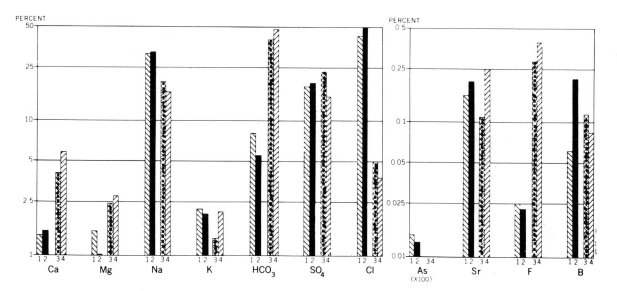

FIG. 11. Bar graphs showing composition of waters from four different areas. Water entering west side of Death Valley (sample 2) is chemically like water at Mesquite Flat (sample 1) northwest of salt pan. Water entering east side of Death Valley (sample 3) is like water at Ash Meadows in Amargosa Desert east of Death Valley (sample 4). Water from the two directions is very different. That from the west is chloride sulfate water low in calcium and fluoride but containing some arsenic; that from the east is bicarbonate sulfate water high in calcium and fluoride and containing no arsenic. Each bar represents average of several analyses. (From U.S. Geol. Survey Prof. Paper 494-B.)

quantity of water being discharged. The fluorine content of Death Valley drinking water is, incidentally, excessive for children, but water without fluorine is available without charge.

The similarity of these two waters, in the Amargosa Desert and in Furnace Creek, is made more striking by their contrast with the water discharging into Death Valley at the big springs at Cottonball Marsh on the west side of Cottonball Basin. The water in these springs is chemically like the water in Mesquite Flat and apparently discharges via the faults along the southwest side of Salt Creek Hills. Figure 11 illustrates the similarity of the

Cottonball Marsh and Mesquite Flat waters and their dissimilarity to the water at Furnace Creek. Whereas the Amargosa Desert—Furnace Creek water contains chiefly bicarbonate sulfate, the Cottonball Marsh—Mesquite Flat water contains chiefly chloride sulfate. The former has more magnesium, calcium, and fluorine and less arsenic than the latter.

STREAMS

The hydrologic basin draining to Death Valley covers approximately 9,000 square miles. Altitudes range from 282 feet below sea level on the floor of Death Valley near Badwater to 11,049 at Telescope Peak in the Panamint Range opposite Badwater. The low point on the rim of the basin, about 900 feet above sea level, is at the south in the valley connecting Death Valley and Soda Lake. The Amargosa River, which drains about 6,000 square miles, has the largest drainage basin discharging into the salt pan. The next largest drainage area is that of Salt Creek, draining the northwest arm of Death Valley, about 2,200 square miles. Although Salt Creek has the smaller drainage basin, its geology is such that it discharges more surface water to the salt pan than does the Amargosa.

Most of the time the Amargosa River is a dry wash (figs. 12, 33). The mouth of the river, where it enters the salt pan, is 240 feet below sea level. Along the river are a number of large springs, including Saratoga Springs, and the springs at Tecopa, Shoshone, Eagle Mountain, Ash Meadows, and Beatty. Water from these springs enters the ground and moves toward Death Valley, some of it via the fill along the course of the river and, as already noted, some of it via faults along Furnace Creek. In wet years the river may have a continuous flow, but this happened only once (winter of 1957-58) during my six-year study. At the more headward springs the proportion of sulfates to chlorides

FIG. 12. Amargosa River at Coyote Hole near south end of salt pan; view is northeast and downstream. During 1969 flood this channel, 120 feet wide and 3 feet deep, was filled almost to overflowing; since channel slope is 12 feet per mile, peak discharge must have been approximately 750 cubic feet per second or 1,500 acre-feet per day. Downstream, where river turns right, channel is in an arroyo about 10 feet deep, which may have resulted from arching of salt pan by uplift at Mormon Point.

dissolved in the water is about 2:1; where the river enters the salt pan the proportions are reversed.

Salt Creek, which enters the salt pan from the north (figs. 13, 34), drains only a third as much area as does the Amargosa River, and the drainage basin is equally dry, yet Salt Creek enters the salt pan at 210 feet below sea level as a perennial stream with waterfalls and even fish. The seeming anomaly arises from the fact that Salt Creek must cross the impermeable beds of the **Furnace Creek Formation, which was domed to form the Salt**

Creek Hills (fig. 14). Water draining underground from Mesquite Flat in the northwest arm is ponded against these impermeable formations and the overflow forms Salt Creek. There is no stream in Mesquite Flat.

Water in Salt Creek is strikingly different in quality from that in the Amargosa River. At Valley Springs on the Amargosa River, where the underflow is brought to the surface by a barrier of impermeable beds, like the barrier at the Salt Creek Hills, dissolved salts include 37.5 percent carbonates and bicarbonates, 17 percent sulfates, and 45.5 percent chlorides. Salt Creek contains 5.5 percent carbonates, 28.5 percent sulfates, and 66 percent chlorides.

The east slope of the lofty Panamint Range drains to the floor of Death Valley in a series of canyons, each of which discharges onto the apex of a gravel fan 4 to 6 miles long and 1,000 to 1,500 feet high. The total drainage area is about 450 square miles, a third of which is highly permeable ground of the gravel fans. As already pointed out, in today's climate floods coming out of these canyons sink into the gravels and do not cross the fans. Even the 1969 flood discharged into the ground on the fans, although it washed out long sections of roadway on the gravel benches in canyons.

There have been floods, however, of sufficient magnitude to cross the fans. Among these are the floods, probably Pleistocene, which deposited the boulder ridges on Starvation Canyon fan. One flood, probably no more than a few hundred years old, was sufficient to transport logs of pine and juniper 6 feet long and 18 inches in diameter to the toe of Trail Canyon fan (fig. 64).

The rocks in the mountains, though not permeable, have many fractures that provide channels for the collection and discharge of groundwater.

FIG. 13. West distributary of Salt Creek where it crosses smooth, silty rock salt to floodplain in Cottonball Basin. This channel is 32 feet wide and 1 foot deep. Much of the efflorescence on channel upstream from pool is mirabilite (hydrous sodium sulfate).

FIG. 14. Oblique aerial view of Salt Creek crossing Salt Creek Hills, an anticline of Pliocene and Pleistocene formations dividing salt pan (foreground) from structural basin at Mesquite Flat (in distance); view northwest. Light-colored beds (Tfc) in center of anticline are Pliocene Furnace Creek Formation. Dark gravel (Qo) on south flank of anticline and gray gravel on north flank are early Pleistocene and are uplifted less than Furnace Creek Formation. Late Pleistocene gravel (Qy), which forms terraces along tributary of Salt Creek, is domed less than early Pleistocene gravel. (Photo courtesy of John H. Maxson.)

SPRINGS

Springs in the Death Valley area are of four principal kinds. The largest springs, such as Travertine, Texas, and Nevares, are those discharging along the steeply dipping faults of the Furnace Creek fault zone between the Black Mountains and the Funeral Mountains. Other small springs of this type are located along the frontal fault at the foot of the Black Mountains. The large springs, especially those along the Furnace Creek fault zone, have deposited travertine that has accumulated in large mounds (fig. 15). Most of the mounds were formed in Pleistocene time, as suggested by the occurrence of early-type Indian artifacts on their surfaces. Former springs that are now dry may be marked by travertine mounds; one such dry spring, by Furnace Creek Wash, extended its mound over the edge and into the wash (fig. 16). The exposure is to the northeast of Furnace Creek, at the first bend above the inn. The springs discharging from the Furnace Creek fault zone supply water for the permanent residents, a public campground, and the resort establishment; they also irrigate a golf course and a grove of date palms covering about 40 acres.

In the mountains, springs discharge at low-angle faults. Most of the springs in the Panamint Range are of this type; so also is the spring by the highway at Daylight Pass between the Funeral Mountains and the Grapevine Mountains. These springs have water of excellent quality, but they discharge no more than a few gallons per minute.

A third kind of spring occurs where groundwater is ponded by an impermeable structural barrier. The best example (already noted) is the spring zone that maintains the flow of Salt Creek through the Salt Creek Hills, reached by the road along Salt Creek at McLean Spring. Another example may be seen in the hills above the Park Service area, where Pleistocene gravels are thin on the upfaulted impermeable Pliocene formations. Water seeping through the gravels, probably most of it ultimately from Nevares Spring, rests on the impermeable Pliocene beds and

FIG. 15. Travertine mound (white) at Nevares Spring, issuing along a fault at foot of Funeral Mountains. Cambrian Bonanza King Formation is in background.

FIG. 16. Travertine deposits at dry springs along Furnace Creek fault zone. Two levels are discernible: lower one (tl) drapes over side of Furnace Creek Wash (right foreground) and reaches valley floor; upper one (tu) forms bench at skyline in center. Projectile points of prepottery type were found on travertine.

discharges in small springs along the front of the fault escarpment (fig. 17).

Still a fourth type of spring is found around the edge of the salt pan where groundwater is ponded in the gravel and sand of the fans where they grade into silt under the salt pan. Tule Spring, Shorty's Well, Eagle Borax, Bennetts Well, and Salt Spring are examples. Much of the water in these springs is saline, but the quality varies with rock composition at the source and with the amount of recharge from the mountains, a function of altitude and geologic structure.

The water table under the high parts of the gravel fans is many hundreds of feet deep. At the foot of the fans, where the water table is shallow, its slope is 25 to 50 feet per mile. In this zone the salinity of the water increases sharply in short distances toward the salt pan. At Salt Well the salinity is about 0.6 percent; a mile panward it is 4 percent. At Gravel Well the salinity is 0.13 percent; 3,000 feet closer to the salt pan it is 4 percent. The salinity of the groundwater is least under the fans at Hanaupah and Starvation canyons, probably because these fans drain the highest, snowiest, and rainiest part of the Panamints.

From Death Valley Canyon north to Tucki Mountain, most of the mountains are Paleozoic carbonate formations; there springs are small and are charged with carbonates and bicarbonates in solution. South of Death Valley Canyon the rocks are Precambrian and Lower Cambrian quartzite and shale with little carbonate, and there the springs are larger and comparatively high in sulfates and low in chlorides and carbonates.

Very little water enters the salt pan as groundwater from the Black Mountains. Slopes are steep; at the north end the ground is impermeable. Floods off the Black Mountains reach the salt pan and the water there seeps into the ground on the floodplains or evaporates. There are marshes and springs along the faults at the foot of the Black Mountains (fig. 18). The spring at Badwater contains about 3 percent dissolved salts, the same as seawater; there is very little carbonate or bicarbonate, and chlorides are about five times more plentiful than sulfates.

FIG. 17. Springs (marked by vegetation) issuing from base of Pleistocene gravels where they overlie impermeable Pliocene playa beds (white) at fault blocks back of Park Service service area. Foreground is no. 2 gravel.

FIG. 18. Spring along fault zone at foot of Black Mountains which is estimated to discharge 3 to 5 gallons per minute. Most springs thus situated are sulfate springs.

Water discharging at springs at the edge of the salt pan is under a small but distinct hydraulic head. The artesian condition is emphasized by the fact that, at small seeps, discharge increases when the barometer pressure falls and decreases when the pressure rises. This phenomenon is quite noticeable on the floodplain. Earth tides may also be responsible for some of the fluctuations in discharge at these springs.

Discharge at the springs is also affected by changes in the rate of evaporation. At Salt Creek, for example, the flow in winter is two to three times heavier than the flow in late spring. The pool at Badwater which covered about half an acre in 1959 covers noticeably less area in summer than in winter, but it does not go dry. At both places—Salt Creek and Badwater—the recharge is by groundwater; fluctuations in the amount of water at the surface seem to be owing chiefly to differences in evaporation.

ESTIMATING DISCHARGE

Assuming that the evaporation rate along the floor of Death Valley is roughly equivalent to that at the weather station at Park Service Headquarters, about 150 inches annually, the minimum discharge at seeps and springs can be estimated by measuring the extent of the water surface. Any estimate is minimal because some water may seep into the ground, though the loss around the salt pan probably is not large because the ground is not very permeable. Spring discharge at Badwater, for example, is estimated at 5 gallons per minute and, assuming an equal amount of undetected flow, total discharge is estimated at about 10 gallons per minute.

The most extensive marsh on the Death Valley salt pan and the one having the heaviest discharge is Cottonball Marsh at the east foot of Tucki Mountain and on the west side of Cottonball Basin. In its perennial pools of water the desert fish cyprinodont lives (fig. 19). The marsh consists of two areas; the smaller one, covering about 140 acres in 1960, lies half a mile north of the larger one, which covered approximately 500 acres in 1960. In January 1970, owing to wetter weather, both areas were considerably enlarged. Most of Cottonball Marsh is a barren salt- and gypsum-encrusted area with scattered clumps of pickleweed and salt grass. In the winters during our survey we found that 40 percent of the marsh area was water surface; the rest was a wet crust of salt or gypsum. By late spring only 5 to 10 percent of the area

FIG. 19. This pool at Cottonball Marsh, 12 to 18 inches deep, is somewhat saltier than seawater (total salts about 4 percent). Small, minnowlike fish (cyprinodonts) live here. Chemically this water is like groundwater under Mesquite Flat; evidently such pools are maintained by discharge of groundwater along faults intersecting the high water table ponded by uplift at Salt Creek Hills.

was water surface and 90 percent or more was wet salt or gypsum. The discharge was estimated to be 700 gallons per minute. Since the catchment area for the marsh is inadequate to supply this water, the recharge is almost certainly attributable to groundwater discharging along faults from Mesquite Flat.

Another large marsh lies on the opposite side of Cottonball Basin and extends about 3.5 miles along the east side of the salt pan. It is easily accessible from the old dirt road below the present highway. Its southern limit is about 1.5 miles north of the Park Service area. The marsh has a large number of small seeps. There are no big pools, but a thin film of water may be developed on the wet mud surface by tapping the ground. The combined flow in thirty channels was estimated to be 150 gallons per minute, and the loss from 250 acres of salt and mud flat was estimated at 75 gallons per minute; the total discharge is therefore about 225 gallons per minute.

Other springs and seeps along the sides of the salt pan are small. (For the location of springs in Death Valley, see fig. 20.) The springs along the east side are salt springs, many of them twice as salty as seawater. The springs on the west side, though, from Gravel Well to Tule Spring, yield potable water. Total discharge from springs and seeps along the east side of Badwater Basin is estimated at about 70 gallons per minute; discharge from those along the west side is estimated at about 2,850 gallons per minute. The fortyfold increase reflects the greater height and the larger area of mountains discharging to the west side. Because of the heavier discharge the water is of better quality; the salts are flushed panward.

The big springs along the Furnace Creek fault zone—Nevares, Texas, and Travertine—together yield an estimated 2,550 gallons of water per minute. Travertine Spring is much the largest of these.

Altogether the groundwater discharged into the Death Valley salt pan totals about 7,500 gallons per minute, and another 500 gallons per minute is contributed by Salt Creek. An undeter-

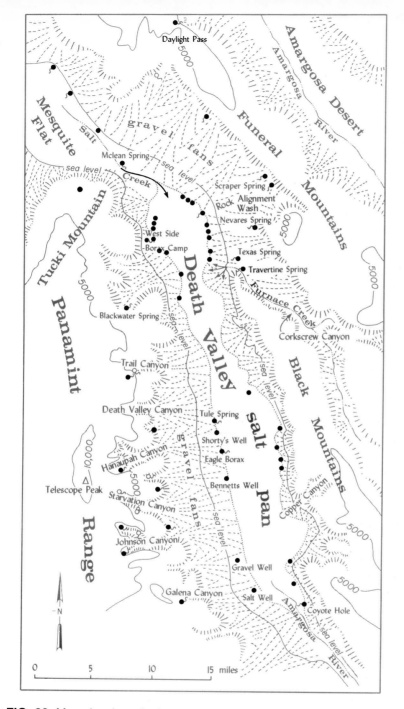

FIG. 20. Map showing principal springs and seeps around salt pan and some of larger ones on gravel fans and in mountains.

FIG. 21. Stands of mesquite around Death Valley salt pan are dying, partly because of cutting but mainly because water table has been dropping and groundwater has become increasingly saline. This scene is along Salt Creek east of Salt Creek Hills.

mined amount enters as underflow from the Amargosa River. These estimates do not include the occasional rare floods that reach the valley via Salt Creek, Furnace Creek Wash (now diverted to Gower Gulch), and the Amargosa River. Floods can make the normally dry channels on the floodplain look like respectable streams, and the floodwater collects in broad pools in the three basins.

If present all at the same time, the total amount of water entering Death Valley in an average year would cover the floodplain part of the salt pan to a depth of only a few inches; obviously the 1969 flood represented extremely unusual conditions of runoff. The evaporation rate is 150 inches a year, but evaporation stops when no water is left to evaporate, which is most of the time. Thus the Death Valley water budget balances, and there is no deficit spending.

The water economy was very different during the middle Holocene, when water stood 30 feet deep across the floor of Death Valley. The discharge into the valley was heavier and the evaporation was less. Evidence of the heavier discharge lies in the now dry springs, the alluviation along the Amargosa River, and the strandline of the lake. That the water supply is continuing to diminish is suggested by the fact that stands of mesquite, which in Death Valley are dependent on groundwater of fairly good quality, have been dying and are not being replaced (figs. 21, 156) by new growths. Whether the decline of mesquite is attributable to a decrease in the supply of water or to an increase in its salinity, or to both factors, is uncertain.

3   The Salt Pan: Orderliness in the Natural Environment

The floor of Death Valley (figs. 22, 23) is a vast evaporating dish covering more than 200 square miles. It is crusted over with a variety of salts (table 1) distributed in zones, which are orderly both horizontally and vertically and which faithfully reflect the differences in solubility of the salts. Under the salt crust, which averages perhaps 2 to 3 feet in thickness, are 25 to 50 feet of silt and clay.

SALT MINERALS—THEIR ZONING AND GEOLOGIC HISTORY

The geologic history of the salt pan, and the crystallization of its salts, can be compared with the evaporation of brine in a dish (fig. 24). Evaporation of the brine yields an orderly sequence of salts, reflecting their differences in solubility. As the brine evaporates, the first precipitates are carbonates of calcium and magnesium ($CaCO_3$, $MgCO_3$), which are deposited around the edge and across the bottom of the dish. As evaporation continues,

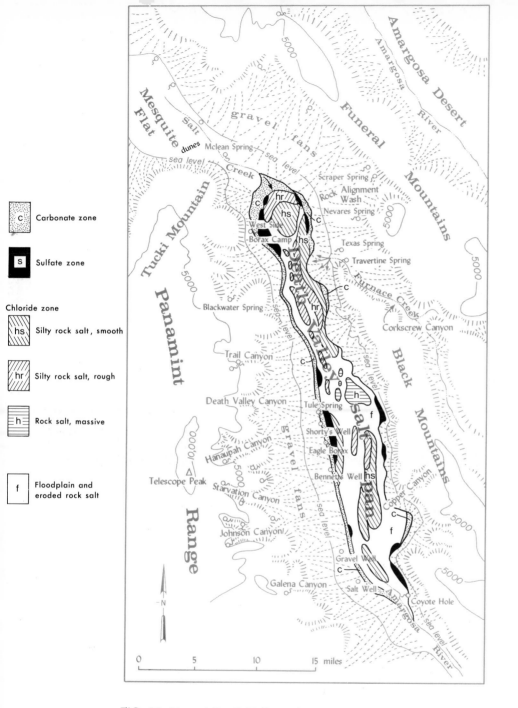

FIG. 22. Map of Death Valley salt pan.

FIG. 23. Concentric rings on Death Valley salt pan, northwest from Dante's View. Badwater is at lower left corner. Gypsum in sulfate zone forms crescentic deposit in lower left center; rock salt of chloride zone extends over several square miles in left center. Shoreline features of Holocene lake which deposited the salts are plainly seen cutting across toes of fans on far side of valley.

Table 1. COMPOSITION AND OCCURRENCE OF SALT MINERALS IN DEATH VALLEY

Mineral	Composition	Known or probable occurrence
te	$NaCl$	Principal constituent of chloride zone and of salt-impregnated sulfate and carbonate deposits.
ite	KCl	With halite.
colite	$NaHCO_3$	Not yet identified; might be found in wintertime as efflorescence on trona or thermonatrite in carbonate zone in Cottonball Basin.
na	$Na_3H(CO_3)_2 \cdot 2H_2O$	Carbonate zone of Cottonball Basin, especially in marshes.
rmonatrite	$Na_2CO_3 \cdot H_2O$	Questionably present on floodplain in Badwater Basin; would be expected in marshes of carbonate zone in Cottonball Basin.
on	$Na_2CO_3 \cdot 10H_2O$	Not yet identified but may be expected, especially in winter, immediately following rains or periods of high discharge at marshes in carbonate zone in Cottonball Basin.
sonite	$Na_2Ca(CO_3)_2 \cdot 2H_2O$	Not yet identified; may be expected in environments where gaylussite would be dehydrated.
lussite	$Na_2Ca(CO_3)_2 \cdot 5H_2O$	Carbonate zone and floodplain in Badwater Basin.
ite	$CaCO_3$	Occurs as clastic grains in sediments underlying salt pan and as sharply terminated crystals in clay fraction of carbonate zone and in sediments underlying sulfate zone.
nesite	$MgCO_3$	Obtained in artificially evaporated brines from Death Valley; not yet identified in salt pan; may be expected in carbonate zone of Cottonball Basin.
omite	$CaMg(CO_3)_2$	Identified only as a detrital mineral; may be expected in carbonate zone.
thupite	$Na_3MgCl(CO_3)$.	An isotropic mineral, having index of refraction in range of northupite and tychite, has been observed in saline facies of sulfate zone in Cottonball Basin.
d (or) hite	$Na_6Mg_2(SO_4) \cdot (CO_3)_4$	
keite	$Na_6CO_3(SO_4)_2$	Sulfate zone in Cottonball Basin.
nardite	Na_2SO_4	Common in all zones in Cottonball Basin and in sulfate marshes in Middle and Badwater basins.
abilite	$Na_2SO_4 \cdot 10H_2O$	Occurs on floodplains in Cottonball Basin immediately following winter storms.
uberite	$Na_2Ca(SO_4)_2$	Common on floodplains except in central part of Badwater Basin; sulfate zone in Cottonball Basin.
ydrite	$CaSO_4$	As layer capping massive gypsum 1 mile north of Badwater. Possibly also as dry-period efflorescence on floodplains.
sanite	$2CaSO_4 \cdot H_2O$	As layer capping massive gypsum along west side of Badwater Basin and as dry-period efflorescence in floodplains.
psum	$CaSO_4 \cdot 2H_2O$	In sulfate caliche layer in carbonate zone, particularly in Middle and Badwater basins; in sulfate marshes and as massive deposits in sulfate zone.
xahydrite	$MgSO_4 \cdot 6H_2O$	Not yet identified but might be expected as dehydration product of epsomite in chloride zone or on floodplains.
omite	$MgSO_4 \cdot 7H_2O$	Not yet identified; probably will be found as efflorescence on floodplains following storms or floods; would dehydrate to hexahydrite during dry periods.
edite	$Na_2Mg(SO_4)_2 \cdot 4H_2O$.	Questionably present in efflorescence on floodplain in chloride zone.
yhalite	$K_2Ca_2Mg(SO_4)_4 \cdot 2H_2O$	Questionably present on floodplain in chloride zone.
ite	$BaSO_4$	Not yet identified but probably will be found in carbonate zone and as clastic grains in sediments underlying salt pan.
estite	$SrSO_4$	Found with massive gypsum.
airerite	$Na_3(SO_4)(F,Cl)$	Not yet identified; might be expected in Cottonball Basin or east side of Middle Basin.
fohalite	$Na_6ClF(SO_4)_2$	Not yet identified; might be expected in Cottonball Basin or east side of Middle Basin.

Table 1   (Continued)

Mineral	Composition	Known or probable occurrence
Kernite	$Na_2B_4O_7 \cdot 4H_2O$	Possibly present in Middle Basin in surface layer of layered sulfate and chloride salts.
Tincalconite	$Na_2B_4O_7 \cdot 5H_2O$	Probably occurs as dehydration product of borax.
Borax	$Na_2B_4O_7 \cdot 10H_2O$	Floodplains and marshes in Cottonball Basin.
Inyoite	$Ca_2B_6O_{11} \cdot 13H_2O$	Questionably present (X-ray determination but unsatisfactory) in floodplain in Badwater Basin.
Meyerhofferite	$Ca_2B_6O_{11} \cdot 7H_2O$	Found in all zones in Badwater Basin and in rough silty rock salt in Cottonball Basin.
Colemanite	$Ca_2B_6O_{11} \cdot 5H_2O$	Questionably present (X-ray determination but unsatisfactory) in floodplain in Badwater Basin.
Ulexite	$NaCaB_5O_9 \cdot 8H_2O$	Common on floodplain in Cottonball Basin; known as "cottonball."
Probertite	$NaCaB_5O_9 \cdot 5H_2O$	A fibrous borate with index of refraction higher than ulexite occurs on dry areas in Cottonball Basin following hot dry spells and in surface layer of smooth silty rock salt.
Soda niter	$NaNO_3$	Weak but positive chemical tests obtained locally.

SOURCE: U.S. Geological Survey Professional Paper 494-B, p. B49.

sulfates of calcium and sodium ($CaSO_4$, Na_2SO_4) are deposited. The proportion of these sulfates depends on the proportion of calcium to sodium. When calcium is plentiful, enough may be left over from the carbonate stage of precipitation to form gypsum, or calcium sulfate ($CaSO_4 \cdot 2H_2O$). When there is only a little calcium, the sulfate deposited is thenardite (Na_2SO_4), and some sodium carbonate (Na_2CO_3) may be deposited. Finally, when the brine reaches maximum salinity, chlorides, mostly sodium chloride or ordinary salt (NaCl), are deposited. Small quantities of the highly soluble magnesium sulfate and of calcium, magnesium and potassium chlorides also appear as the last of the brine evaporates.

The Death Valley salt pan consists of similar concentric zones of salts formed by the evaporation of the Holocene lake (p. 14), but two complexities must be introduced to continue the analogy with the laboratory experiment. First, as evaporation progressed, the floor of Death Valey was tilted eastward. The southern half of the salt pan was crowded against the foot of the Black Mountains and the salt rings were disposed asymmetrically, as illustrated in figure 24, *B*. Similar asymmetry in the salt crust on Great Salt Lake Desert is also attributed to tilting of that valley floor. The second complexity is illustrated in figure 24, *C*. During and since

FIG. 24. Diagram illustrating zoning of salts and history of salt pan. As brine evaporates the least soluble salts, carbonates (*c*), precipitate first. As evaporation continues and salinity increases, sulfates (*s*) are precipitated. Finally, when maximum salinity is reached, the most soluble salts, chlorides (*h*), are precipitated. The salts are rearranged around parts of the valley which are subject to flooding (*f*). (From U.S. Geol. Survey Prof. Paper 494-B.)

evaporation and tilting, new additions of fresh water by rains and floods reworked the salts already deposited. Salts washed from the areas subject to flooding were redeposited around the areas where ponds formed, as happened conspicuously during the 1969 flood.

The salts in the salt pan were deposited in several different hydrologic environments. First, massive amounts of rock salt were deposited at ponded water. On the playa surface around the water, evaporation of groundwater deposited silty rock salt that was washed smooth. Around the edge of the playa was a zone of springs where massive gypsum was deposited, and beyond that was a zone where the water table stood 5 to 10 feet higher than it does today, as indicated by a sulfate caliche layer in the carbonate zone.

Not only are the salts in orderly concentric zones on the surface of the salt pan but they also occur in orderly layers that reflect differences in solubility. As groundwater rises to the surface its salinity increases. Carbonates are deposited at depth; sulfates are deposited in a layer above them; and chlorides form a crust at the surface. If surface water sinks in from above, the reverse may be true.

Superimposed on the concentric zoning is another kind of zoning with respect to the degree of hydration of the different minerals, that is, their content of chemically combined water (table 2). The lesser hydrates (dehydrated minerals) occur in two kinds of environment. They occur at the surface on ground protected against flooding, where dehydration is probably due to the extremely high ground temperatures. They also occur in wet environments, where temperatures are moderate owing to the chilling that results from evaporation, but where salinity exceeds 10 percent. Apparently the salts become dehydrated at moderate temperatures in the presence of sodium chloride.

Variations in the chemistry and the mineralogy of the deposits around the salt pan also depend on the sources of water and sediments coming into the pan. In the north, for example, the

Table 2. NINE SERIES OF HYDRATED SALTS THAT
MAY OCCUR IN DEATH VALLEY

No.	Greater hydrate		Lesser hydrate
1.	$Na_2CO_3 \cdot 10H_2O$ (natron)	$\rightarrow$	$Na_2CO_3 \cdot H_2O$ (thermonatrite)
2.	$Na_2Ca(CO_3)_2 \cdot 5H_2O$ (gaylussite)	$\rightarrow$	$Na_2Ca(CO_3)_2 \cdot 2H_2O$ (pirssonite)
3.	$CaSO_4 \cdot 2H_2O$ (gypsum)	$\rightarrow 2CaSO_4 \cdot H_2O$ (bassanite) $\rightarrow$	$CaSO_4$ (anhydrite)
4.	$Na_2SO_4 \cdot 10H_2O$ (mirabilite)	$\rightarrow$	Na_2SO_4 (thenardite)
5.	$Na_2Ca(SO_4) \cdot 4H_2O$ (wattevilleite)	$\rightarrow$	$Na_2Ca(SO_4)_2$ (glauberite)
6.	$MgSO_4 \cdot 7H_2O$ (epsomite)	$\rightarrow MgSO_4 \cdot 6H_2O$ (hexahydrite) $\rightarrow$	$MgSO_4 \cdot H_2O$ (kieserite)
7.	$Na_2B_4O_7 \cdot 10H_2O$ (borax)	$\rightarrow Na_2B_4O_7 \cdot 5H_2O$ (tincalconite) $\rightarrow$	$Na_2B_4O_7 \cdot 4H_2O$ (kernite)
8.	$Ca_2B_6O_{11} \cdot 13H_2O$ (inyoite)	$\rightarrow Ca_2B_6O_{11} \cdot 7H_2O$ (meyerhofferite) $\rightarrow$	$Ca_2B_6O_{11} \cdot 5H_2O$ (colemanite)
9.	$NaCaB_5O_9 \cdot 8H_2O$ (ulexite)	$\rightarrow$	$NaCaB_5O_9 \cdot 5H_2O$ (probertite)

brines contain little calcium, and the sodium there appears both as carbonate and as sulfate. In the southern part of the pan, where calcium is plentiful, it appears both as carbonate and as sulfate, and sodium is restricted to the chloride zone.

In brief, although precipitation of salts is controlled largely by relative solubilities, other factors affect the process by modifying the environment. Examples are temperature of the solution including its range of temperature, rate of evaporation of the solution, suspended and dissolved matter in the liquid, and turbulence of the liquid.

There are about a hundred borate, nitrate, sulfate, carbonate, and chloride salts of the alkalies (sodium and potassium) and of alkaline earths (magnesium, calcium, strontium, and barium). Each salt has its special uses, and each has a history all its own.

Table 3. CHEMICAL RELATIONSHIPS AMONG COMMON SALT, SALT CAKE, BLEACHING POWDER, SODA ASH, CAUSTIC SODA, AND WASHING SODA, AS ILLUSTRATED BY THE LEBLANC PROCESS

1. The first step is the decomposition of ordinary salt by sulfuric acid to produce salt cake and hydrochloric acid:

$$2\ NaCl + H_2SO_4 \longrightarrow Na_2SO_4 + 2\ HCl$$

2. Chlorine is obtained from the acid by oxidizing it with a peroxide:

$$MnO_2 + 4\ HCl \longrightarrow MnCl_2 + 2\ Cl + 2\ H_2O$$

3. Bleaching powder is obtained by passing the chlorine across calcined limestone:

$$CaCO_3 + heat \longrightarrow CaO + CO_2$$

$$CaO + Cl_2 \longrightarrow CaOCl_2$$

4. The salt cake obtained in step 1 may be used for manufacturing soda carbonate (soda ash) by heating it with coal and limestone:

$$Na_2SO_4 + CaCO_3 + 2C \longrightarrow Na_2CO_3 + 2CO_2 + CaS$$

5. Caustic soda in turn can be obtained from soda ash and slaked lime:

$$Na_2CO_3 + Ca(OH)_2 \longrightarrow 2NaOH + CaCO_3$$

6. Soda ash dissolved in water produces washing soda, $Na_2CO_3 \cdot 10H_2O$

About a dozen of the minerals are common in the Death Valley salt pan. The chemistry of some of these and their relationship to one another is illustrated by one of the earliest and simplest processes for manufacturing synthetic minerals, the Leblanc process (table 3).

Of the salts in Death Valley, the borates have aroused the greatest interest and have had the most economic importance. From 1907 to about 1925, the Death Valley area was the principal source of borax and boric acid consumed in the United States. Before the discovery of borates in California and Nevada, the

principal world sources were Tibet, Tuscany (Italy), and Iquique
(Chile). The earliest source was Tibet, where for hundreds of
years borax was used for making fine glaze and for soldering
gold. Marco Polo is credited with introducing borax to Europe
about A.D. 1275.

Chloride Zone   In the central part of the salt pan lies the
chloride zone; covering about three-quarters of the valley floor, it
is by far the most extensive of the salt deposits. Around the
chloride zone is the sulfate zone, a discontinuous zone surrounding the pan. The carbonate zone, when present, is on the outer
edge of the salt pan.

Four facies of the chloride zone have been distinguished. At the
center is massive rock salt (fig. 25), 2 to 6 feet thick and covering

FIG. 25. Massive rock salt in central part of chloride zone is clean salt with
little admixed silt. The salt, 2 to 6 feet thick, is polygonally fractured; one
fracture can be seen at right center extending left from under handle of
hammer. Surface is irregular, with salt hummocks projecting a foot upward;
hummocks are further roughened by lacy but tough and spiny salt growths,
sharp enough to cut leather. Openings that reach down to wet mud
underlying salt are marked by efflorescences that form delicate structures of
varying shapes; one seen here (opposite tip of hammer) is a cylindrical
growth with hemispherical top.

7 or 8 square miles. It is nearly pure rock salt containing less than 0.5 percent insoluble residue; it can be seen at the parking area at the Salt Pools.

Surrounding the massive rock salt is a facies of rough silty rock salt (fig. 26), deposited where surface water was ephemeral while the Holocene lake dried; the ground was alternately flooded and desiccated, as the lake rose and fell; when it fell, salts were deposited by evaporation of groundwater and mud was incorporated in the salt. Cracks in the salt layer extend upward through the silt, but desiccation cracks in the silt end downward at the salt. The deposit, about 3 feet thick, has a very rough upper surface which is mantled by silt; it can be seen beside the road across the valley floor at the Devils Golf Course.

Surrounding the rough silty rock salt is a smooth facies of silty rock salt, most easily seen west of the road at Mushroom Rock. At the surface is a layer of silt up to about 6 inches thick. Under it is a layer of rock salt 6 to 12 inches thick. Under the salt is mud to a depth of 25 to 50 feet. The difference in form between the smooth and the rough facies is attributed to variations in the frequency of washing by surface water. The rough facies, deposited by groundwater, was infrequently flooded; the smooth facies developed in areas where the surface was frequently washed, as opposite the toe of Furnace Creek fan (fig. 27). In the silty rock facies, the salt layer contains about 15 percent insoluble residue; the silt layer contains about 35 percent salt. Both layers contain 5 percent sulfate salts.

In addition to these three facies, there are areas where the silty rock salt has been subject to a very recent washing, as by the 1969 flood, and the salt crust there has been eroded. Although not shown on figure 22, these washed places were mapped separately as eroded rock salt, the fourth facies of the chloride zone. The extent of the erosion in 1969 can be measured by the changes in the limits of the eroded rock salt as shown on the geologic map accompanying the Death Valley report from which this account is abstracted (U.S. Geol. Survey Prof. Paper 494-B,

FIG. 26. The Devils Golf Course, crossed by road to west side of Death Valley, is composed of rough, silty rock salt. The siltfree, massive rock salt in central part of chloride zone (fig. 25) was probably deposited in standing water as Holocene lake gradually dried. The silty rock salt shown here forms peripheral zone that presumably was a wet mud flat much of the time; mud was incorporated in salts as they crystallized.

CARBONATE ZONE → | ← SULFATE ZONE → | ← FLOOD PLAIN

Anhydrite

Blister crust of rock salt

Nodular sulfate layer

Massive gypsum

Rock salt

Recent alluvium

Calcitic sandy silt

Calcitic sandy silt; locally with trona

FIG. 27. Smooth, silty rock salt opposite foot of Furnace Creek fan. It has been subject to frequent washing by water running off the fan, as shown by rounded cobbles of highly porous, lightweight, vesicular lavas washed down onto surface.

FIG. 28. Diagrammatic section of massive gypsum deposit. Gypsum is 2 to 5 feet thick; uppermost layer (6 inches thick) is dehydrated to bassanite or anhydrite. Under gypsum is calcitic sandy silt. Where gypsum ends panward at a wash, as in this section, rock salt is deposited on silty sand and rises into gypsum in irregular veins, producing a rough surface analogous to rough, silty rock salt. Relationship between massive gypsum and nodular sulfate layer in carbonate zone is not known. (From U.S. Geol. Survey Prof. Paper 494-B.)

pl. 1). Some effects of the erosion can be seen beside the road across the valley floor just west of the Devils Golf Course.

Sulfate Zone   In the sulfate zone, major differences in mineralogy reflect differences in the sources of the brines. At the north end of Death Valley there is little calcium, whereas there is enough sodium to combine even with the carbonate, yielding trona ($Na_3H(CO_3)_2.2H_2O$) and thermonatrite ($Na_2CO_3.H_2O$). The sodium sulfate, thenardite (Na_2SO_4), is found in the sulfate zone in the north. In the southern part of Death Valley, however, there is much calcium. Sulfate deposits there are gypsum, or hydrous calcium sulfate ($CaSO_4.2H_2$); there is no sodium carbonate in the carbonate zone of this part of the valley. This difference in mineralogy may be seen by comparing the deposits at the edge of the salt pan below the Park Service residential area with those near Badwater.

The mineralogy of the borates similarly reflects these differences. In the north, where sodium predominates, common minerals are borax ($Na_2B_4O_7$, $10H_2O$) and ulexite ($NaCaB_5O_9.H_2O$); in the south, where calcium predominates, meyerhofferite ($Ca_2B_6O_{11}.7H_2O$) is common.

The gypsum deposits, 2 to 5 feet thick, rest on mud (fig. 28). One of the accessible deposits is a mile north of Badwater and 1,000 feet west of the highway. Capping the gypsum is a 6-inch layer of anhydrite ($CaSO_4$) or of the hemihydrate, bassanite ($2CaSO_4.H_2O$) (fig. 29). On the east side of Death Valley the cap rock is anhydrite; on the west side it is bassanite. Ground temperatures may be higher where there is anhydrite, owing possibly to reflection of the afternoon sun from the precipitous front of the Black Mountains. Where the gypsum ends against a floodplain, water seeps laterally under the gypsum and deposits rock salt in irregular layers which disrupt the surface by heaving—the wick effect (fig. 28)—as may be seen half a mile east of Tule Spring. Except for such introduced rock salt the gypsum contains few salts.

Near these gypsum deposits, spring-fed sulfate marshes where gypsum is being deposited today provide clues as to how the massive gypsum was formed. The sulfates grow in lumpy cauliflowerlike forms 8 to 10 inches in diameter (fig. 30); the crusts and the interiors are of very different composition. The soluble sodium chloride is deposited in the crust; the less soluble gypsum forms granules in the interior. Further, there is a striking difference in the composition of the water in these marshes in wet and dry years. In wet years, when water flows out of the marshes, the amount and the proportion of sodium chloride are

FIG. 29. Layering near surface of massive gypsum shown on wall of a pit dug a foot into deposit. Capping surface is an inch of silt, probably deposited by wind and containing a wide variety of minerals. Below silt is a firm white layer, 4 to 6 inches thick, of anhydrous calcium sulfate (anhydrite or bassanite), and below it is hydrous calcium sulfate (gypsum), which resembles bread crumbs.

low. In dry years the chloride to sulfate ratio is 15 to 1; in wet years it is only 3 to 1. At the time of the middle Holocene lake, discharge at the springs was heavier than it is today, perhaps heavy enough to keep sodium chloride flushed out of the system where calcium sulfate was being precipitated.

In other words, gypsum deposits are believed to have been formed at springs larger than those existing today, after the middle Holocene lake had dropped below the level where the deposits occur. The gypsum could hardly have been flooded by that lake or it would have become impregnated with sodium chloride, as did older deposits at and above this level.

FIG. 30. Lumpy, mammillary growths of gypsum in spring-fed sulfate marsh. Growths reach 8 to 10 inches in diameter. Interiors are composed of damp granular gypsum; walls or crusts are cemented with sodium chloride. In wet seasons the chloride may be largely flushed out of marsh and transported in solution to central part of valley; in dry seasons the chloride is deposited on the gypsum. Good examples may be seen in marshes just north of Badwater.

Carbonate Zone   The carbonate zone, at the edge of the salt pan, is mostly sand or silt (fig. 31), but it contains tiny crystals of calcite ($CaCO_3$). These well-formed fragile calcite crystals were not transported with the sand; they formed in the sand. Although calcite is the characteristic mineral of the carbonate zone, it is a minor constituent. Actually there is more sulfate and in places more chloride than carbonate in what is called the carbonate zone. Where sodium is abundant, as in the northern part of Death Valley, sodium carbonates as well as calcite are found.

Salts in the sandy facies of the carbonate zone occur in layers. The surface layer, 3 to 20 inches thick, contains calcite or sodium carbonate. Beneath the surface layer is a layer 1 to 6 inches thick which contains sulfate salts. In the northern part of the valley these salts are mostly thenardite, or sodium sulfate; in the southern part, they are mostly gypsum, or calcium sulfate. Underlying the sulfate layer is sand or silt containing carbonate and chloride salts.

FIG. 31. Detail of ground surface in carbonate zone. Surface is sand, loosely cemented by salts, including carbonates. Surface is pitted by water seeping into ground rather than rilled by runoff.

The sand is probably late Pleistocene in age, but the layering of the salts is younger. Panward the sand grades into silt, and in this belt groundwater reaches the surface or comes sufficiently near it to cause salts to be deposited at the surface, forming a chloride layer of rock salt in blisterlike growths (fig. 32). This layer is underlain by silt at the top of which is a layer with nodules of sulfate minerals; the silt contains the tiny, well-formed crystals of calcite ($CaCO_3$). The hummocks of salt are ephemeral and collapse with a soft crunch when stepped upon. As a matter of fact, each salt zone has a characteristic crunch.

FIG. 32. Silt facies of carbonate zone which, subject to frequent wetting by rise of groundwater, develops crust, mostly sodium chloride, in blisterlike growths an inch or so thick. Silt under crust contains nodules of sulfate minerals; below them is another layer of silt, containing carbonates.

Floodplain   The parts of the valley floor which are subject to seasonal flooding are referred to as floodplain (fig. 7). Parts of the floodplain in Cottonball Basin which frequently are flooded evaporate to bare mud flats speckled with borates in the form of small pellets (fig. 109). Known as cotton balls, the pellets are composed of ulexite, or sodium calcium borate. They were the object of the early borax work on the floor of the valley, dating from the 20-mule-team operations of 1882.

The Brines   The Amargosa River, where it enters Death Valley, is usually dry (fig. 33), but it can discharge floods. By the time floods reach the central part of the salt pan they are saturated with sodium chloride, and a salt crust develops. In places, the crust formed giant saucerlike growths (fig. 36). It is easy to

FIG. 33. Channel of Amargosa River where it enters salt pan. View is south and upstream to Shoreline Butte. When this channel overflowed in 1969 flood, the water spread between mounds of sand on floodplain.

visualize how continued growth of these saucers would form the salt crust that characterizes the rough silty rock salt in the chloride zone. Ground subject to frequent flooding has similar, but smaller, salt crusts.

The composition of water in the Amargosa River changes along the 50-mile stretch upstream from the edge of the salt pan. Total dissolved solids decrease upstream from 2 percent at the edge of the salt pan to 0.1 percent about 50 miles upstream. As total salts decrease upstream, the proportion of carbonates increases and the proportion of chlorides decreases.

PATTERNED GROUND

Polar and alpine regions are characterized by ground patterns consisting of polygons, circles, small terraces, and stripes of rocks. The mechanisms that produce these features are not fully understood, but frost action is clearly the major factor. Ground patterns over layers of salt resemble those over layers of ice; those in Death Valley are formed by solution and recrystallization of salts. Much of the ground on and around the Death Valley salt pan is hardened with salts, like the permanently frozen ground of the north country; the valley can count on having a white Christmas, though the white is a bit salty. Since ground patterns of the two regions are similar, salt can be thought of as a warm form of ice.

The simplest forms of ground patterns are the familiar cracks caused by the drying and shrinking of muds (figs. 34, 62). The spacing of the cracks and the width and depth of the openings depend on the quantity of salts and the size of salt crystals in the ground. Cracks in the ground can become permanent, or nearly permanent, by deposition and crystallization of salt in them.

The cracks narrow downward, and the salt forming in them is wedge shaped. Not only does growth of the wedge widen the crack but the wedge also grows upward (fig. 35). When such ground is flooded again, it forms a network of natural salt pans, and salt crust is deposited across flat areas between ridges. In time these areas may become tremendous saucerlike forms (fig. 36).

Ground patterns on the floodplain differ from one place to another in an orderly way, reflecting wetness or frequency of flooding. Before the 1969 flood the lowest places, like those west of Badwater and in the central part of Middle Basin, rarely, if ever, completely dried. The water is saturated with salt, and a salt crust 1 to 12 inches thick which forms the surface has a complex pattern of polygons and superimposed nets or hum-

FIG. 34. Desiccation cracks along Salt Creek opposite Furnace Creek fan. In the middle a pool of water dried leaving thin white crust of salt, with cracks 1 to 2 feet apart and as much as an inch wide. Along each side is a belt without a salt crust and with cracks 6 to 8 inches apart. At each side of wash the ground is still wet and has no cracks. All the belts are silty clay, but center one has the most salts (5 percent by volume); they form mesh of crystals (glauberite-thenardite-halite) up to 3 millimeters in diameter. The bordering belt, with more closely spaced cracks, has 4 percent salts, with crystals 1 to 2 millimeters in diameter. The wet ground without cracks contains less than 3.5 percent salts, with crystals half a millimeter or less in diameter.

FIG. 35. As wedges of salt deposited in polygonal cracks grow upward (and probably downward) the cracks widen. The wedges, or salt ridges, enclose miniature mud flats which trap silt and brine; they become crusted with salt as brine evaporates. The originally flat salt crust may thus grow upward and become a thick slab of salt segmented by vertical, polygonal columns.

FIG. 36. Salt saucers, final stage of salt wedges and polygonally cracked salt crusts. Secondary cracks have so widened that crusts have been thrust outward. Cracks photographed here were old enough to have collected some silt (dark surface in lower left), and bottom of each saucer is pierced by one or more drain holes. These saucers, a mile west of Badwater, were dissolved during 1969 flood.

mocks (fig. 37). The washes tributary to these low places frequently are washed by small floods entering the valley, and salt
crusts along them are thin and ephemeral. When buried by silt,
as in much of the chloride zone of the salt pan, such crusts retain
their polygonal patterns. The capping silt develops its own
pattern of cracks independent of the patterns in the underlying
salt.

Open salt pools 30 inches deep, associated with circle patterns,
develop on the floodplain wherever salt layers in the mud have
dissolved and the surface has collapsed (fig. 38). Such pools may
be seen on the floodplain west of the end of the road at the Salt
Pools, north of Badwater. There, too, can be seen the polygonal
cracks in the thick layer of massive rock salt forming the central
part of the chloride zone of the salt pan.

The cracks in massive rock salt are old features. At the Salt
Pools part of the rock salt was leveled for a road in 1942.
Pinnacles that characterize the weathered surface of the massive
rock salt were removed, but the cracks remain. The only discernible change in approximately thirty-five years is a roughening of
about a quarter of the surface by the development of delicate
knifelike ridges of salt an inch or so high separated by rounded
grooves, all trending north. Very likely the cracks in these thick
layers of salt formed as a result of changes in volume caused by
temperature changes. On a warm day one can hear the cracking
as a series of sharp and melodious pings.

The growth of these salt structures is due in large part to the
capillary rise of water in underlying muds. Salt forms under
objects resting on the mud and in time raises them above the
surface (fig. 39). They look like the rock-capped pinnacles that
form on the surface of glaciers, where exposed ice has melted
except in the shade beneath a rock. In Death Valley, however, the
salt pinnacles grew upward. The capillary rise of salt is also
illustrated by the bursting of wooden posts set in wet saline
muds (fig. 40).

FIG. 37. Parts of floodplain which are perennially damp or wet develop hummocky, blisterlike growths of silty salt. A recent flood that spread onto this ground deposited new salt (white) in depressions, but there was not enough water to destroy blisterlike hummocks.

FIG. 38. Circle pattern formed by collapse of salty mud into pool of salty water. Such structures are common on floodplain in vicinity of salt pools.

silt and granular gypsum   bassanite

FIG. 39. Twigs, stones, and other debris resting on damp mud of valley floor are lifted upward by salts crystallizing beneath them.

FIG. 40. Wood post shattered by salt is an example of evaporation effects on salt pan. When this post, 3.5 feet long, was set in floodplain of Salt Creek, about 1910, the ground level was at lower edge of bulge. Water rising in the wood evaporated at ground level and burst the wood with precipitated salts.

FIG. 41. Ground patterns on massive gypsum. Surface layer of silt and granular gypsum, about 1 inch thick, develops desiccation cracks which end downward at a firm layer of bassanite. As silt layer continues to dry, it becomes loose fluff and the cracks are lost. The bassanite is divided into irregular polygons by cracks that extend upward into silt layer and end downward at crumbly gypsum underneath bassanite.

Polygonal cracks marked by ridges looking very much like those in figure 35 also form in wet marshy sulfate ground where there is a flat crust of gypsum. Dry massive gypsum is also cracked polygonally. For example, just east of Tule Spring 3 to 5 feet of massive but crumbly and porous gypsum has a firm cap rock 2 to 4 inches thick of bassanite or anhydrite (see table 1). About an inch of granular calcium sulfate mixed with silt lies on this cap rock (see fig. 41).

After a wetting, the surface layer cracks polygonally into primary plates 4 to 5 inches wide; with continued drying these plates divide into small ones 1 to 2 inches wide. When completely dry, the surface becomes fluffy and is eroded by wind; open cracks fill up, the polygonal pattern is destroyed, and microdunes develop.

The primary crack pattern in the silty surface layer faithfully follows a crack pattern in the underlying layer of cap rock; the cracks extend downward into the crumbly gypsum. The cap rock of bassanite or anhydrite on gypsum is thickest along cracks. Between the cracks, in the interiors of the polygons, the cap rock may be reduced to irregular veins of bassanite or anhydrite. The secondary cracks in the silty surface layer clearly are desiccation cracks, but the primary ones and the cracks in the cap rock seem to be related to volume changes attributable to hydration or dehydration of calcium sulfate.

A different kind of polygonal crack in massive gypsum may be seen along the shore of the Holocene lake on the east side of the valley just north of the road to the Devils Golf Course (fig. 42). The ground that was submerged is saline and has a hummocky salt crust (fig. 32). When the lake existed, the water table extended at lake level into the hill, and in the capillary fringe above the table a layer of rock salt was formed. This layer has cracked polygonally and the cracks are marked at the surface by depressions, where stones tend to collect. Polygonal ground patterns marked by stones like these are a common feature in ground that is permanently frozen or subject to much freeze and

FIG. 42. Polygonal patterns in gypsiferous ground at shoreline of Holocene lake, just north of road crossing valley at Devils Golf Course. Polygons are formed by cracks in a layer of rock salt 6 to 8 inches below surface; the salt is at least 3 inches thick, and perhaps much thicker than that. It formed at capillary fringe above water table, which extended into the hill when the lake existed. Surface cracks in salt are marked by troughs in which stones have collected.

FIG. 43. Hillsides on Tertiary playa beds in Mustard Canyon area are crusted with salt 6 inches or more thick and are cracked into polygonal slabs up to 10 feet in diameter. Cracks are not vertical; they are perpendicular to the slope.

thaw. This Death Valley location shows they may be formed by salt as well as by ice.

Salt crusts have also formed over steep hillsides of Tertiary formations (fig. 43) in the vicinity of Mustard Canyon, which is readily accessible. The crusts were presumably formed by salts being leached from the old playa beds. Good examples may be seen of crusts that are cracked polygonally. The polygons are 10 feet in diameter; the salt crust is 1 or 2 feet thick. In other places the crust on these formations has developed stripes. There are stripes in the mountains too, but they are natural levees produced by flash floods and are not attributable to salt.

A different form of patterned ground, marked by small steplike terraces, is found on hillsides of the oldest gravel formation, no. 2 gravel, and on the desert pavement on the surface of the gravel. On slopes steeper than 10 degrees the terraces are small lobate structures 3 to 12 inches high, 1 to 4 feet wide across the contour, and as much as 15 feet long parallel to the contour. The treads, which slope 3 to 11 degrees, consist of fine gravel; the risers in front slope at angles as steep as 37 degrees (fig. 44).

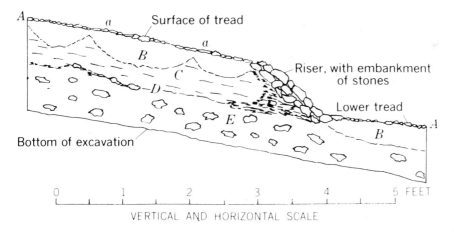

FIG. 44. Section of terracette on hillside of no. 2 gravel. *A.* Pavement of fine pebbles on tread of terracette. *a.* Locations of salt efflorescence on tread. *B.* Dry, porous, light-gray layer cemented by salt; some stones. *C.* Damp, soft, brown layer with reddish spots, which rises to within 0.5 inch of surface under salt efflorescence. *D.* Firmly cemented stony layer. *E.* Parent gravel, not well cemented. (From U.S. Geol. Survey Prof. Paper 494-B.)

These terracettes in Death Valley seem to be old features that are now stable. Rows of dowels set across them (fig. 45) showed no downhill movement in four years, and trails abandoned more than fifty years ago exhibit no sign of being disturbed by the downhill creep of the terracettes. The terracettes probably originated in an environment with somewhat more moisture and with more frequent wetting and drying than now. Such condition prevailed at the time of the Holocene lake that produced the salt pan.

THE SALT PAN—AN OUTDOOR MUSEUM

Information gleaned from the study of the Death Valley salt pan provides an unusually satisfying example of orderliness in the natural environment. The salt pan is an outdoor museum of the mineralogy and geochemistry of salt deposits. It serves also as an educational medium, reminding observers of the importance of salt in both past and present.

The part played by salt in the life of primitive people must remain largely a matter of conjecture, because salt they may have stored has been lost. One theory is that the demand for salt developed as a result of changing from nomadic to agricultural living. Those who ate raw or roasted meats needed no salt, but salt may have become a necessity when meats were boiled and supplemented by cereals and vegetables.

Salt came into demand for use as a preservative, and perhaps it was the eating of fish and meat preserved in salt which led to the habit of using it for seasoning. In the *Iliad* (IX, 254; ca. 800 B.C.), Homer speaks of the use of salt for seasoning: ". . . then sprinkled o'er the meat with salt." In Homer's time salt seems to have been commonplace, at least among seafaring people; the *Odyssey* states (XVII): " . . . would'st not give even so much as a grain of salt to thy suppliant." Evidently salt was not then in universal use, however, for the *Odyssey* also tells of other peoples (XI) who "know not the sea, neither eat meat savoured with salt."

In the fifth century B.C. the early Greek historian Herodotus wrote about the salt pans at the Dneiper River where "salt forms in great plenty about its mouth without human aid" (IV, 53). Near Susa, capital of the Persian king, Darius, was a well that "yields produce of three different kinds . . . bitumen, salt, and oil" (VI, 119). This account is reminiscent of the occurrences of salt

FIG. 45. Since dowels set on terracettes showed virtually no movement in several years, terracettes seem to be stable in present climate. Probably they formed when Death Valley was wetter, as it was at time of Holocene lake.

and oil in New York and Pennsylvania, first developed by the Indians and later leading to America's first oil well. Xerxes' route through Phrygia passed "a lake from which salt is gathered" (VII, 29). Herodotus also mentions salt springs, salt deposits, salt quarries, and houses of salt in the Libyan deserts (IV).

Death Valley has some houses, or small shelters, likewise made of salt. One is a mile southeast of the West Side Borax Camp; another is close to Salt Creek at the old crossing directly west of Furnace Creek Ranch. The litter around these houses suggests that they were built before 1900, probably during the early days when borax was produced from the salt pan.

Passages about salt in both the Old and the New Testament reveal its importance in religious ceremonies of the Israelites: "And every oblation of thy meat offering shalt thou season with salt" (Lev. 2:13); "a covenant of salt" (Num. 18:19); "And thou shalt offer them before the Lord, and the priests shall cast salt upon them" (Ezek. 43:24); "every sacrifice shall be salted with salt" (Mark 9:49). Other passages indicate the high favor in which salt was held; "Ye are the salt of the earth" (Matt. 5:13); "Salt is good: but if salt have lost its savour, wherewith shall it be seasoned!" (Luke 14:34; see also Matt. 5:13; Mark 9:50); "Let your speech be always with grace, seasoned with salt" (Col. 4:6).

The use of salt as money has been preserved in the English vocabulary. In the days of the Roman republic soldiers were first paid in salt; later their compensation was changed to a *salarium* (salary), or an allowance of money in lieu of salt. Perhaps the best summary of salt is given by Ephraim Chamber's *Cyclopaedia* (1752): "The great property of salt is, that it is incapable of corruption."

4 Gravel Fans

The fans separating the mountains from the salt pan include the
driest ground in Death Valley. On the fans precipitation is less
than in the mountains and very little more than on the valley
floor. The gravel fans are highly permeable, and water that runs
onto them quickly seeps into the ground. Only rare floods rising
in the mountains flow very far onto the gravels. Fans built of
muds from Tertiary formations are highly impermeable, and
water quickly runs off them.

The fans are Quaternary in age, but here and there older
underlying formations protrude through them. The fans record
the downsinking of the valley during Quaternary time (the past 2
or 3 million years) and the faulting and tilting that accompanied
the sinking. They record stages when lakes existed in Death
Valley, as well as an earlier period when the valley was open and
drained southward to the Colorado River.

How did the rocks become fractured to supply the vast quantity
of fragmented material found as boulders, blocks, pebbles, and
sand in the fans? Present rates of weathering and shattering of
rocks in the mountains seem inadequate to have provided the
gravels in Quaternary time. Probably most of the gravel was
produced during those stages of the Pleistocene when more

northerly regions became glaciated. There was no glaciation in
the Panamint Range or in the other mountains around Death
Valley; the mountains were probably cold enough but there was
not enough precipitation. To produce glaciers, not only cold but
also precipitation is required.

No doubt there was more snow on the mountains then than
now. Meltwater would fill cracks and other openings in the rocks
and freezing and thawing would shatter the rocks. Moreover, the
rapid melting of large snowbanks could produce floods of a
magnitude not seen today. The history of the occurrence of lakes
in Death Valley provides compelling evidence that in times past
water was vastly more abundant than it is now.

DIFFERENCES IN FANS

Gravel fans sloping to Death Valley may be considered in four
groups. The largest, both in area and in volume, are those along
the east foot of the Panamint Range (see p. xii and fig. 46). These
fans are 5 to 6 miles long, and their surfaces rise from below sea
level, where they merge with the salt pan, to more than 1,000 feet
above sea level. The area of the Panamint Range draining to
these fans is about twice the area of the fans. The rocks in the
mountains are mostly Precambrian and Paleozoic sedimentary
formations with small areas of granite and volcanic rock (chap.
5). Geophysical surveys (gravity and magnetic measurements)
indicate that the bedrock floor is 6,000 feet under the fans; the
deepest part is near the west edge of the salt pan and roughly
opposite Telescope Peak. These fans are readily accessible by all
except the most modern low-slung cars; roads climb the fans at
Trail, Hanaupah, Johnson, and Warm Spring canyons.

A second group of fans rises from the salt pan to the Funeral
Range. These fans are as long and as high as those rising to the
Panamint Range, but the fan form is partly obscured by hills and

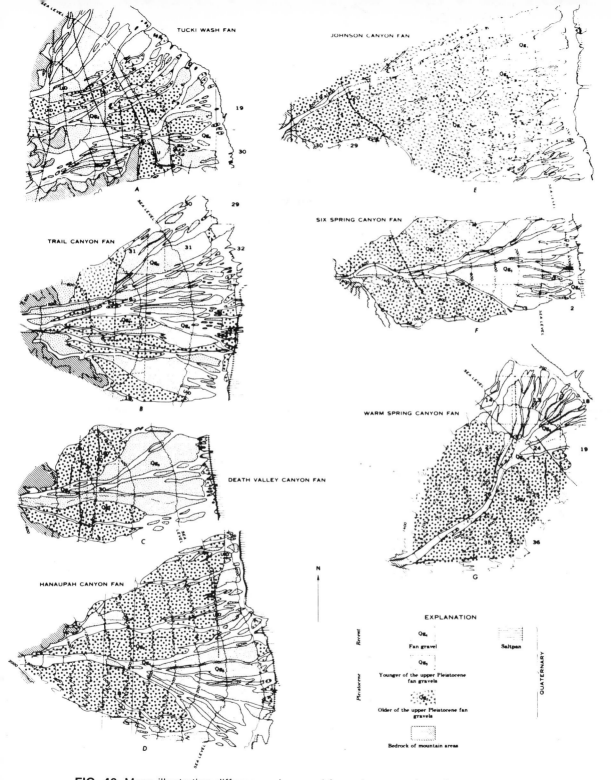

FIG. 46. Maps illustrating differences in gravel fans along east foot of Panamint Range, arranged from north (Tucki Wash fan) to south (Six Spring Canyon fan). (From U.S. Geol. Survey Prof. Paper 494-A.)

ridges of older rocks protruding through the gravels (fig. 47), as along the road to Beatty. Also, these gravel deposits are on the average much thinner than those along the foot of the Panamint Range, and their volume is correspondingly less. The area of the Funeral Mountains draining to those fans is about the same as the area of the fans. This ratio of mountain area to fan area is half that along the east side of the Panamint Range and may account for the difference in the fans. The parent rocks in the mountains are not very different; those in the northern part of the Funeral Range are mostly Precambrian whereas those in the south are Paleozoic.

A third group of fans comprises those along the foot of the Black Mountains on the east side of the salt pan south of Badwater. The east side highway runs along the foot of these fans, which are small (fig. 48), probably because the floor of Death Valley has been tilting eastward during Quaternary time (see p. 120 and chap. 6). The fans have been sinking with the valley floor and their lower ends are buried by overlap of playa sediments.

The fourth group of fans are those at the foot of the Black Mountains along the highway north of Badwater. They are composed of fine-grained sediments derived by erosion of Tertiary formations in the northern part of the Black Mountains. The area of these fans is about equal to the area of the mountains draining to them, but the mountain summits are only 2,000 to 4,000 feet high.

FAN PATTERNS

Differences in fan patterns (fig. 49) reflect differences in the structural history of mountain fronts. Some fans are short and the fronts are fan-based (fig. 49, *A*) where there has been maximum valley sinking, as along the front of the Black Mountains south of

FIG. 47. Gravel fans sloping from Funeral Mountains (in distance) are marked by ridges of upfaulted gravels. Ridge seen here extends north from Park Service residential area; faulting that formed it predated Pleistocene lake that produced horizontal beach terraces on gravelly hillsides. The terraces, up to 250 feet above sea level, are 500 feet above floor of Death Valley; the Pleistocene lake must have been that deep. Fault block is formed of Funeral fanglomerate (cf. fig. 86); foreground is no. 3 gravel.

FIG. 48. Fans along faulted steep front of Black Mountains, seen here, are smaller than those along west side of Death Valley because they have small catchment areas. Moreover, because east side is being tilted downward, its broad fans (like the one in immediate foreground) extend beneath fine-grained sediments of playa floor. Most fans show several shades of staining by iron and manganese oxides (desert varnish). The most stable and oldest surfaces are darkly stained: the least stable and youngest surfaces show no staining.

Badwater and at the east foot of Tucki Mountain. As already noted, the short fans in these areas are partly buried by playa deposits; the fans are probably larger than they seem to be at the surface.

The long fans sloping down from the Panamint Range vary in an orderly way northward along the foot of the mountains. Toward the south the fans extend a short distance into the canyons (fig. 49, *B*). Northward the mountain front is frayed (fig. 49, *C*), and at the north end the fans extend far into the mountains along major canyons and are wrapped round hills of bedrock (fig. 49, *D*). The Panamint Range seems to have been

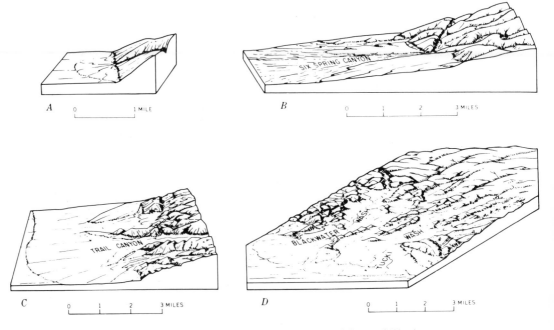

FIG. 49. Fan patterns in Death Valley. *A.* Fan-based front of Black Mountains at Coffin Canyon; view north. *B.* Fan-bayed east front of Panamint Range at Six Spring Canyon; view south. *C.* Fan-frayed east foot of Panamint Range at Trail Canyon; view south. *D.* Fan-wrapped east foot of Panamint Range at Blackwater and Tucki washes; view southwest. (From U.S. Geol. Survey Prof. Paper 494-A.)

tilted northward while the fans were forming. The fan-wrapped east foot of the Panamint Range at Blackwater and Tucki washes can be seen very well from the Furnace Creek fan.

AGES OF GRAVEL

On the gravel fans derived from Precambrian or Paleozoic rocks the gravels of different ages form different kinds of ground, each having a distinctive drainage pattern (fig. 50).

The oldest gravels that still preserve the fan form rest unconformably on faulted and tilted blocks of the cemented gravel of the Funeral Formation; they are not themselves much faulted.

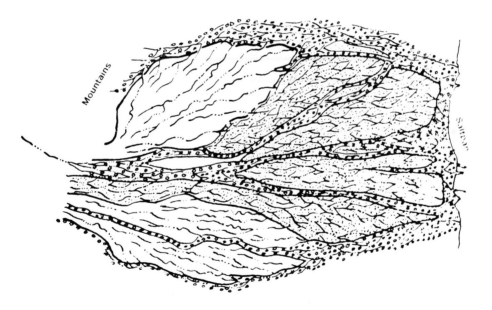

FIG. 50. Map illustrating differences in drainage pattern on older and younger gravels. On no. 2 gravel (white areas) drainage is parallel, tending toward dendritic. On younger gravels, no. 3 (stippled areas) and no. 4 (circle pattern), drainage is braided. (From U.S. Geol. Survey Prof. Paper 494-A.)

These gravels are known as no. 2 gravel, the Funeral Formation (see p. 116) being no. 1. No. 2 gravel lacks calcite veins; some layers are cemented with calcium carbonate, as in the Funeral Formation, but the cement is neither so thick nor so strong. Both have a smooth surface called desert pavement (see below) composed of angular blocks, slabs, and flakes disintegrated from the once rounded stream gravels that formed the deposits. No. 2 gravel can be distinguished from younger gravels in at least six ways: (1) no. 2 gravel deposits form the highest benches on the upper parts of the fans; (2) no. 2 gravel extends under and is overlapped and buried by younger gravel on the lower parts of the fans; (3) no. 2 gravel on any particular fan is more bouldery than younger gravels on the same fan; (4) no. 2 gravel is more cemented than younger gravels; (5) the surface of no. 2 gravel is smooth desert pavement, whereas surfaces on younger gravels are rough (fig. 53); (6) stream-worn cobbles and boulders on the surface of no. 2 gravel have disintegrated to form a new crop of angular rock fragments (fig. 51).

The disintegration of stones on no. 2 gravel is most advanced where the gravels extend into the zone having abundant salts, a feature that can be seen beside the highway 3 miles north of Badwater and beside the westside highway along the foot of Trail Canyon fan. The effectiveness of salts in promoting rock disintegration is illustrated very clearly by the disintegration of concrete at bench marks and of stones at graves on salty ground. This phenomenon can be seen at the cement monuments recording bench marks beside the road along the west side of the valley.

The youngest gravel, no. 4, clearly is still being washed by floods, for it is found along present-day washes (see left side of fig. 52). The stones are rounded and firm and have little or no stain left on them. The ground is loose and difficult to cross even with four-wheel-drive vehicles.

The surface of the intermediate gravel, no. 3, is less rough than the surface of no. 4, but it is very much rougher than the surface of no. 2 gravel. The surface of no. 3 gravel is young enough so

FIG. 51. Boulder of granitic rock (quartz monzonite) in process of breaking up by exfoliation (peeling of rock surface as skins are peeled from an onion). The exfoliation is probably caused by hydration, which makes some of the minerals swell. The exfoliated shells break down further into coarse grit, seen at base of rock.

FIG. 52. Contrast between no. 3 gravel (right) and no. 4 gravel (left). No. 3 gravels are stained with desert varnish; no. 4 gravels, presently subject to washing, are not stained. No. 3 gravels are firmly embedded in the ground, whereas no. 4 gravels are loose. No. 3 gravel shown here lacks vegetation because, being more compacted than no. 4, less water infiltrates the ground.

that the levees of coarse cobbles and small boulders form ridges
1 to 2 feet high. These can be rough even for a four-wheel-drive
car, but the bottoms of the washes, which are about as wide as
the levees, are smoothly floored with desert pavement (figs. 52,
53). The stones are firm; that is, they are not fragmented like
those on no. 2 gravel. On older no. 3 surfaces the stones may
have a weathered rind up to half an inch thick; on younger
surfaces there is no rind. Also, the older surfaces are darkly
stained with desert varnish (see p. 83), whereas the younger
gravels lack the stain. The differences between these gravels are
well shown along the Hanaupah Canyon road a mile southwest
of Shorty's Well.

FIG. 53. Three different kinds of gravel surfaces are seen in this view,
looking east from highway 190 to Funeral Mountains. In foreground is no. 3
gravel, with boulders forming natural levees between gravelly washes.
Boulders are firmly embedded in the ground; washes between them are
floored with desert pavement. The younger no. 4 gravel (light color, at left) is
composed of loose stones. An older gravel (no. 2) forms dark-surfaced fan
(center) with its smooth surface of desert pavement. The ridge, formed of
Funeral Fanglomerate, is older than Pleistocene beach terraces impressed
against it.

There are three types of surface on no. 3 gravel. One type has not been subject to washing recently and is only a little less smooth than the adjacent older gravels. Such surfaces are somewhat rough because they contain ill-sorted small boulders and cobbles. Surfaces of the second type, subject to flooding but not recently, have levees of small boulders along the sides of washes floored with pebbles, and both the washes and the levees, as noted above, are darkly stained with desert varnish. The varnish is thicker and darker on the first two types of surface than on any other gravel deposits in this part of Death Valley.

The youngest surfaces on no. 3 gravel, of the third type, grade into no. 4. On these parts of the fans the levees have retained their stain, but the washes between them have been subject to recent washing and are not stained. The surfaces of no. 2 gravel grade into no. 3, for we are dealing with a continuous process. The major differences between no. 2 and no. 3 can be distinguished, but the gravels are not everywhere sharply separated, as brought out by a few additional examples.

Much of the ground covered by no. 3 gravel actually has only a thin veneer of gravel spread over an eroded surface of no. 2. No. 3 gravel is neither so thick nor so extensive as the underlying no. 2.

Except on the lower parts of the fans the surface of no. 3 gravel is lower than that of no. 2 and generally less than 10 feet above no. 4. But no. 3 gravel overlaps the lower edges of no. 2 gravel and at such places it has accumulated in small fans on top of it. In the same way, no. 4 has accumulated locally on the lower edges of no. 3 (fig. 54). In other words, the older gravels slope more steeply than the younger ones (except at faults) and their lower edges extend under and are overlapped by the younger gravels, because of tilting of the older gravels. At the mouth of Death Valley Canyon faults displace no. 2 gravel but not the younger gravels.

Around Death Valley, the surfaces of the Funeral Formation and of no. 2 gravel are without vegetation. No. 3 gravel has sparse vegetation, but no. 4 can have considerable vegetation. (For vegetation, see chap. 10.)

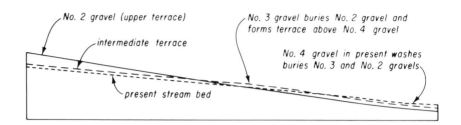

FIG. 54. Diagrammatic profile of fans along west side of Death Valley, illustrating downfan shift in position and overlap of younger gravels on older gravels. No. 4 gravel overlaps no. 3 gravel below point where no. 3 gravel overlaps no. 2 gravel. (From U.S. Geol. Survey Prof. Paper 494-A.)

Solution pits like those in limestone country (karst topography) are common where Tertiary formations are overlain by no. 2 gravel. They may be seen in the East Coleman Hills. A large pit, which is not easily accessible, is on the west side of the valley north of Hanaupah Canyon. Referred to as "the crater," it is 15 feet deep and 50 feet in diameter. The bottom of the pit is 60 feet higher than the wash along the south side of the gravel; the pit probably was formed by a solution of calcium carbonate cementing the base of the gravel.

KINDS AND SIZES OF STONES

The stones forming the gravels reflect the formations in the mountains from which the gravels were derived, and the size of the gravels in part depends on the kind of rocks. On Trail Canyon fan the gravels contain about equal proportions of quartzite and carbonate rocks, with only minor amounts of igneous and metamorphic rocks. Most of the large boulders are about 2 feet in diameter, although some are as much as 6 feet. On Johnson Canyon fan the gravel is about 80 percent quartzite, 10 percent granitic rocks, and 10 percent carbonate rocks; the boulders are small like those on Trail Canyon fan.

Hanaupah and Starvation Canyon fans, however, which are 20 percent granite rock, 60 percent quartzite, and 20 percent carbonates and argillites, drain from Hanaupah Canyon granite southeast of Telescope Peak. These fans contain many larger boulders (fig. 51), many of them more than 10 feet in diameter; some measure as much as 30 feet. The boulders are distributed along the entire fan, down to at least 250 feet below sea level. They were probably deposited before Death Valley sank below sea level and still had exterior drainage. No matter what process is postulated for transporting the boulders to their sites, vast quantities of water would be required. With so much water, there should have been a lake, but had there been a lake the bouldery deposits would have built deltas, not fans. Coarse deposits in so large a volume pose no problem if Death Valley was higher and had exterior drainage to the south at the time they were deposited.

A former channel of Furnace Creek filled with no. 2 gravel is well exposed along Furnace Creek Wash (fig. 55). It can be seen

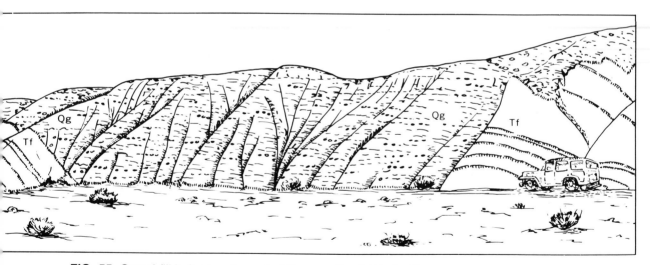

FIG. 55. Gravel fill in former channel of Furnace Creek Wash. Channel is eroded in Furnace Creek Formation (Tf), which dips steeply northeast (right); bottom is below level of present wash. The fill (Qg) is about 50 feet thick. View is north in tributary to Furnace Creek Wash opposite Zabriskie Point. (From U.S. Geol. Survey Prof. Paper 494-A.)

from the highway opposite Zabriskie Point. The gravel fill here contains almost 90 percent carbonate rocks, whereas fanglomerate of the Funeral Formation in that area contains a high proportion of quartzite. Apparently the principal source of gravel fill was in the southern part of the Funeral Mountains, whereas the fanglomerate in the Funeral Formation must have been derived in large part from the northern end of the Black Mountains. Furnace Creek has cut downward 50 feet since the gravel was deposited.

FAULTING ACROSS THE MOUTH OF FURNACE CREEK WASH

The 50 feet of downcutting by Furnace Creek which followed deposition of this channel gravel was probably caused by structural uplift of the wash relative to the present Furnace Creek fan. The surface of the fill, projected, would extend about 50 feet above the fan. This uplift is probably the one that raised the gravel terraces just north and south of the mouth of Furnace Creek (fig. 56) and produced the hanging valleys along the foot of the Black Mountains farther south (fig. 57).

DESERT PAVEMENT

The smooth surface of the gravel fans is known as desert pavement; it may be seen beside roads that cross no. 2 gravel on the fans. The pavement consists of a single layer of closely spaced stones (fig. 58) over a layer of sand and silt having open pores, or vessels (fig. 59), and resting on the gravels forming the fans. It is likely that the stone-free silt was blown onto the fans by winds. Freezing and thawing of the silt when moist, or solution and crystallization of salts in the silt, could push the stones upward to the surface while the finer material settled under them, just as shaking a mixture of sand and pebbles causes the pebbles to rise

FIG. 56. Terraces eroded on playa beds of Furnace Creek Formation and mantled by fan gravels have been uplifted 75 feet by movement on fault (buried under gravel) along front on Black Mountains, just south of mouth of Furnace Creek Wash; view southeast.

FIG. 57. Hanging valley at Gower Gulch at front of Black Mountains. The broad U-shaped valley was eroded to level of gravel fan but was uplifted 75 to 100 feet by movement on fault along front of mountains. The gulch is now incised in a narrow gorge at bottom of uplifted valley.

FIG. 58. Detail of desert pavement on surface of old (no. 2) gravel. Closely spaced angular stones are fragments of large rounded ones that originally composed gravel deposit.

FIG. 59. Silt layer under desert pavement on Death Valley fans may contain 0.5 to 4.0 percent (by volume) of water-soluble salts. The silt underlying the stones is very porous (vesicular) and has been cracked polygonally, apparently because there is enough wetting by rain and dew to make silt swell when wet; later, when it has dried, the silt shrinks.

FIG. 60. Weathering of desert pavement progresses from below as well as from above. Underside of limestone pebble (bottom center) has been etched by solution, evidently of acid. The environment is generally alkaline, but during moist periods the population of organisms under the stones may be sufficiently large to produce acid film on undersides.

to the surface. Slabby stones commonly stand vertically and project above the pavement surface. In general, the silt layer is thicker on older fans than on younger ones.

A pebble on the desert pavement is subjected to three very different environments. The upper surface, exposed to high temperatures and maximum temperature change, is being corroded and sandblasted. On the sides of the pebbles is a narrow band where temperatures are less extreme and where there is maximum wetting by dew as well as by surface water. This narrow zone has a considerable population of microorganisms and even some megascopic ones, algae. The undersides of the pebbles experience moderate temperatures and condensation of soil moisture; the latter is sufficient to pit the undersides of carbonate rocks (fig. 60). The underlying silt is cracked polygonally and vesiculated because it shrinks and swells as it becomes moistened by dew and then dries (fig. 59).

Desert pavement can form in a very short time. Where the ground consists of pebbles and loose sand, the wind can blow away the finer material, causing the pebbles to settle. They serve as an armor protecting the ground against further wind erosion. These very young deposits, though, do not have a silt layer under the pebble pavement. The silt layer on no. 2 gravel commonly is 6 inches thick; on no. 3 gravel it usually is only about 1 inch thick. Archaeological sites perhaps a thousand years old may have a quarter-inch layer of silt.

DESERT VARNISH—THE BLACK STAIN

Stones on older gravel surfaces are stained black with desert varnish, a coating of iron and manganese oxide. Just how the varnish is deposited is not known, but water is required to transport the iron and manganese in solution. At the present time water seeping over stones or down cliff faces leaves such a stain.

But how did the extensive dry surfaces of gravel fans become stained?

The stain is old and dates from a wetter period. The evidence for this conclusion is archaeological. Sites and artifacts dating from about A.D. 1 are not stained with desert varnish except where they are exposed to seep water, but older sites and artifacts are stained (figs. 112, 114, 124, 131). Just before the beginning of the Christian era, Death Valley was wet enough to contain a lake 30 feet deep, and in late Pleistocene time it was wet enough to contain a lake several hundred feet deep (see chap. 2). During wet periods there must have been enough moisture on the gravel fans to deposit the stain. If the stones were then encrusted with algae or lichens, the organic slime coating the rock surfaces would have hastened oxidation and precipitation of iron and manganese.

Whatever its origin, desert varnish is a useful guide to relative ages of rock surfaces. Young surfaces are not stained; old ones are (figs. 48, 52).

WIND EFFECTS

In addition to contributing to the formation of desert pavement, winds in Death Valley can cause sandblasting. Some boulders and cobbles on no. 3 gravel are smoothed and grooved in this way, and occasionally they develop new facets (fig. 61). Examples may be seen along the south side of Hanaupah Canyon fan, 1 to 1.5 miles due west of Eagle Borax; on a bench at the mouth of the wash at the north end of the Artists Drive fault block, about 500 feet east of the road; and at Salt Creek Hills. In the latter area stone artifacts are etched by sandblasting. Glass bottles that have been exposed to wind are sometimes frosted and etched.

FIG. 61. Boulders smoothed and faceted by wind abrasion (sandblasting) at toe of fan on north side of Artists Drive exit. Wind direction is toward upper right, about parallel to handle of hammer. The rock is lava, and the vesicles were caused by bubbles of steam when lava was still molten. The abrading sand elongated the vesicles.

MUDFLOWS

Fans sloping from the north end of the Black Mountains are unlike those elsewhere around the salt pan, for they are composed of fine-grained sediments (fig. 62). The badlands of Tertiary formations in the northern part of the Black Mountains, such as those that can be seen from Zabriskie Point (fig. 83), erode easily, and floods discharge as mudflows from the canyons. Since the fans are composed of fine-grained sediments, the ground is impermeable and floods commonly discharge all the way to the salt pan. At one point the highway to Badwater was buried under so many feet of mud that it was more economical to lay new highway paving on top of the mudflow after it had dried than to clear the old pavement. The surface of these mudflows may become mud-cracked like the muds on the floor of the valley.

These fans are also unlike gravel fans in being easily eroded. The fan at Gower Gulch has been cut by an arroyo because the discharge from the gulch has been increased by diverting Furnace Creek floodwaters into the head of the gulch near Zabriskie Point. The diversion was made to protect the buildings on Furnace Creek fan against flooding by Furnace Creek Wash.

The fan of no. 2 gravel at Starvation Canyon has three tremendous ridges radiating down the fan and evidently marking old mudflows (fig. 63). The ridges are 2 to 3 miles long, 500 to 1,000 feet wide, and 50 to 75 feet high. In volume they range from 8 to 25 million cubic yards. Each ridge has a narrow crest with a wash alongside it; the sides are mantled with huge boulders and slope evenly to the adjoining fan surfaces. To move these gravels required floods on a scale far exceeding the magnitude of any historic floods in the area. The most satisfactory explanation is that the mudflows date from a Pleistocene glacial period when snow collected for several seasons high in the Panamint Range and the meltwater was rapidly released by a sudden spring thaw. More recent and exceptionally heavy floods off the fans west of the salt pan are recorded by juniper and pine logs washed to the foot of Trail Canyon fan (fig. 64).

The floods that produced these effects were several magnitudes heavier than even the 1969 flood, which washed out the highway in Emigrant Canyon and 19 miles of the highway along the west side of the pan. That storm supplied Death Valley with enough water to permit small boats to cross from Badwater to Eagle Borax. But the permanent effects of the 1969 flood were slight compared with the effects of the floods that produced the fields of huge boulders.

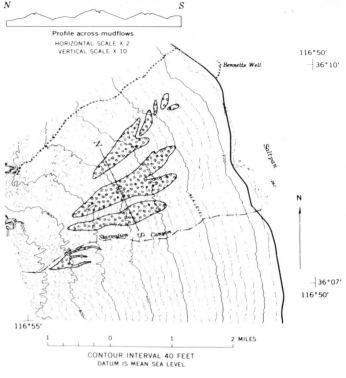

Profile across mudflows
HORIZONTAL SCALE X 2
VERTICAL SCALE X 10

CONTOUR INTERVAL 40 FEET
DATUM IS MEAN SEA LEVEL

FIG. 62. Cracked surface of mudflow on fan at Artists Drive. The cracked silt rests on fine gravel. Silt in foreground is 7 inches thick, with cracks 18 to 24 inches apart; silt in background is 2.5 inches thick, with cracks about 6 inches apart. Cracks tend to form at right angles rather than meeting at 120 degrees and forming hexagons.

FIG. 63. Map and profile across mudflows on Starvation Canyon fan. Topography from U.S. Geological Survey topographic quadrangle; Bennetts Well, 1952. (From U.S. Geol. Survey Prof. Paper 494-A.)

FIG. 64. A major flood off east side of Panamint Range is recorded by juniper and pine logs collected at foot of Trail Canyon fan. Some logs are 6 feet long and 1.5 feet in diameter; they must have been transported to toe of fan from head of canyon. So unusual a flood may have been caused by an accumulation of snow followed by sudden thaw and warm rain.

5 Rocks: A Billion Years of Death Valley History

Nearly every kind, color, and hardness of rock can be found in Death Valley. Igneous rocks once were molten. Like lavas they crystallized from melts, some of them at depth without reaching the surface; these are known as intrusions. Lavas and related volcanic rocks that reached the surface are referred to as extrusions. Sedimentary rocks were originally slushy sediments, like mud. Some of the sediments are marine in origin, deposited at a time when the seas extended across what is now Death Valley; other sediments were deposited on ancient playas and ancient fans rather like the present ones. Metamorphic rocks, more than a billion years old, are the oldest in Death Valley; once sedimentary or igneous, they were recrystallized by deep crustal strains or by heat over a long period of time.

Owing to earth movements, all these rocks have been broken (faulted) by fracturing and contorted (folded) by bending. Although at first sight the rocks seem to be a hopelessly jumbled and haphazard mass, they do have orderliness. This chapter describes the rocks themselves; chapter 6 tells how they became faulted and folded.

TIME SEQUENCE

Death Valley rocks represent all the great eras of geologic time: Precambrian (more than 600 million years ago), Paleozoic (about 600 to 225 million years ago), Mesozoic (225 to 60 million years ago), and Cenozoic (the past 60 million years). Rocks are divided into distinctive units called formations, which are listed and briefly summarized in table 4.

Precambrian rocks exposed in the mountains bordering Death Valley include at least 3,000 feet of metamorphic rocks belonging to the crystalline basement and a sequence of much younger and only slightly metamorphosed sedimentary rocks (Pahrump Series; see table 4) roughly 10,000 feet thick. Overlying the Pahrump and also included in the Precambrian are three sedimentary formations, the Noonday Dolomite, the Johnnie Formation, and the Stirling Quartzite, together about 7,000 feet thick. These Precambrian rocks are mostly shale or siltstone (rocks from muds) and quartzite (a sand rock), but they include a few beds of carbonates or dolomite (precipitates of calcium and magnesium carbonate).

Paleozoic rocks, also of sedimentary origin, are mostly carbonates. They represent all the periods of the Paleozoic from Cambrian to Permian and aggregate about 30,000 feet in thickness, Triassic formations nearby, 8,000 feet thick, include sedimentary carbonate rocks, siltstone, and volcanic rocks.

Metamorphic rocks belonging to the crystalline basement are a sample of the older part of the continental crust of North America. The rocks look impressively massive and strong where they form the bold front of the Black Mountains south of Badwater, but in fact, beginning in late Precambrian time, they were part of a fragile margin of the crust of North America and were repeatedly bent downward to form troughs (geosynclines) many miles deep, scores of miles wide, and hundreds of miles long at the western edge of the continental crust. The site of Death Valley was in the geosynclines; the site of Grand Canyon was on a

Table 4. ROCK FORMATIONS EXPOSED IN DEATH VALLEY AREA

System	Series	Formation	Lithology and thickness	Characteristic fossils
	Holocene		Fan gravel; silt and salt on floor of playa; less than 100 feet thick.	None
Quaternary	Pleistocene		Fan gravel; silt and salt buried under floor of playa; perhaps 2,000 feet thick.	
		Funeral fanglomerate	Cemented fan gravel with interbedded basaltic lavas; gravels cut by veins of calcite (Mexican onyx); perhaps 1,000 feet thick.	Diatoms, pollen.
	Pliocene	Furnace Creek Formation	Cemented gravel, silty and saliferous playa deposits; various salts, especially borates; more than 5,000 feet thick.	Scarce.
	Miocene	Artist Drive Formation	Cemented gravel; playa deposits; much volcanic debris; perhaps 5,000 feet thick.	Scarce.
Tertiary	Oligocene	Titus Canyon Formation	Cemented gravel; mostly stream deposits; 3,000 feet thick.	Vertebrates, titanothere etc.
	Eocene and Paleocene		Granitic intrusions and volcanics; not known to be represented by sedimentary deposits.	
Cretaceous and Jurassic			Not represented; area was being eroded.	
Triassic		Butte Valley Formation of Johnson (1957)	Exposed in Butte Valley 1 mile south of this area; 8,000 feet of metasediments and volcanics.	Ammonites, smooth-she brachiopods, belemnit and hexacorals.
	Pennsylvanian and Permian	Formations at east foot of Tucki Mountain	Conglomerate, limestone, and some shale. Conglomerate contains cobbles of limestone of Mississippian, Pennsylvanian, and Permian age. Limestone and shale contain spherical chert nodules. Abundant fusulinids. Thickness uncertain on account of faulting; estimate 3,000 feet +; top eroded.	Beds with fusulinids, especially *Fusulinella.*
Carboniferous	Mississippian and Pennsylvanian (?)	Rest Spring Shale	Mostly shale, some limestone; abundant spherical chert nodules. Thickness uncertain because of faulting; estimate 750 feet.	None.

Table 4   (Continued)

System	Series	Formation	Lithology and thickness	Characteristic fossils
	Mississippian	Tin Mountain Limestone and younger limestone	Mapped as 1 unit. Tin Mountain Limestone 1,000 feet thick, is black with thin-bedded lower member and thick-bedded upper member. Unnamed limestone formation, 725 feet thick, consists of interbedded chert and limestone in thin beds and in about equal proportions.	Mixed brachiopods, corals, and crinoid stems. *Syringopora* (open-spaced colonies), *Caninia* cf. *C. cornicula*.
Devonian	Middle and Upper Devonian	Lost Burro Formation	Limestone in light and dark beds 1-10 feet thick give striped effect on mountainsides. Two quartzite beds, each about 3 feet thick, near base; numerous sandstone beds 800-1,000 feet above base. Top 200 feet is well-bedded limestone and quartzite. Total thickness uncertain because of faulting; estimated 2,000 feet.	Brachiopods abundant, especially *Spirifer, Cyrtospirifer, Productilla, Carmaro-toechia, Atrypa.* Stromatoporoids. *Syringopora* (closely spaced colonies).
Silurian and Devonian	Silurian and Lower Devonian	Hidden Valley Dolomite	Thick-bedded, fine-grained, and evengrained dolomite; mostly light color. Thickness 300-1,400 feet.	Crinoid stems abundant, including large types. *Favosites.*
Ordovician	Upper Ordovician	Ely Springs Dolomite	Massive black dolomite; 400-800 feet thick.	Streptelasmatid corals: *Grewingkia, Bighornia.* Brachiopods.
	Middle and Upper (?) Ordovician	Eureka Quartzite	Massive quartzite, with thin-bedded quartzite at base and top; 350 feet thick.	None
	Lower and Middle Ordovician	Pogonip Group	Dolomite, with some limestone, at base; shale unit in middle; massive dolomite at top. Thickness, 1,500 feet.	Abundant large gastropods in massive dolomite at top: *Palliseria* and *Maclurites,* associated with *Receptaculites.* In lower beds: *Protopliomerops, Kirkella,* Orthid brachiopods.
	Upper Cambrian	Nopah Formation	Highly fossiliferous shale member 100 feet thick at base; upper 1,200 feet is dolomite in thick alternating black and light bands about 100 feet thick. Total thickness of formation 1,200-1,500 feet.	In upper part, gastropods. In basal 100 feet, trilobite trash beds containing *Elburgis, Pseudagnostus, Homagnostus, Elvinia, Apsotreta.*

Table 4   (Continued)

System	Series	Formation	Lithology and thickness	Characteristic fossils
Cambrian	Middle and Upper Cambrian	Bonanza King Formation	Mostly thick-bedded and massive dark-colored dolomite; thin-bedded limestone member 500 feet thick 1,000 feet below top of formation; 2 brown-weathering shaly units, upper one fossiliferous, about 200 and 500 feet, respectively, below thin-bedded member. Total thickness uncertain because of faulting; estimated about 3,000 feet in Panamint Range; 2,000 feet in Funeral Mountains.	The only fossiliferous is shale below limest member near middle formation. This sha contains linguloid brachiopods and tril trash beds with fragr of *"Ehmaniella."*
	Lower and Middle Cambrian	Carrara Formation	An alternation of shaly and silty members with limestone members; transitional between underlying clastic formations and overlying carbonate ones. Thickness about 1,000 feet but variable because of shearing.	Numerous trilobite tr beds in lower part y fragments of olenell trilobites.
	Lower Cambrian	Zabriskie Quartzite	Quartzite, mostly massive and granulated due to shearing; locally in beds 6 inches to 2 feet thick; not much crossbedded. Thickness more than 150 feet; variable because of shearing.	No fossils.
Cambrian and Cambrian(?)	Lower Cambrian and Lower Cambrian(?)	Wood Canyon Formation	Basal unit is well-bedded quartzite above 1,650 feet thick; shaly unit above this 520 feet thick contains lowest olenellids in section; top unit of dolomite and quartzite 400 feet thick.	A few scattered olenellid trilobites a archaeocyathids in upper part of forma *Scolithus?* tubes.
Precambrian		Stirling Quartzite	Well-bedded quartzite in beds 1-5 feet thick comprising thick members of quartzite 700-800 feet thick separated by 500 feet of purple shale; crossbedding conspicuous in quartzite. Maximum thickness about 2,000 feet.	None.
		Johnnie Formation	Mostly shale, in part olive brown, in part purple. Basal member 400 feet thick is interbedded dolomite and quartzite with pebble conglomerate. Locally, tan dolomite near middle and at top. Thickness more than 4,000 feet.	None.
		Noonday Dolomite	In southern Panamint Range, dolomite in indistinct beds; lower part cream colored, upper part gray. Thickness 800 feet. Farther north, where mapped as Noonday(?) Dolomite, contains much limestone, tan and white, and some limestone conglomerate. Thickness about 1,000 feet.	*Scolithus?* tubes.

Table 4   (Continued)

System	Series	Formation	Lithology and thickness	Characteristic fossils
		Unconformity		
		Kingston Peak(?) Formation	Mostly conglomerate, quartzite, and shale; some limestone and dolomite near middle. At least 3,000 feet thick. Although tentatively assigned to Kingston Peak Formation, similar rocks along west side of Panamint Range have been identified as Kingston Peak.	None.
Precambrian	Pahrump Series	Beck Spring Dolomite	Not mapped; outcrops are to the west. Blue-gray cherty dolomite; thickness estimated about 500 feet. Identification uncertain.	None
		Crystal Spring Formation	Recognized only in Galena Canyon and south. Total thickness about 2,000 feet. Consists of basal conglomerate overlain by quartzite that grades upward into purple shale and thinly bedded dolomite; upper part, thick-bedded dolomite, diabase, and chert. Talc deposits where diabase intrudes dolomite.	None.
		Unconformity		
		Rocks of the crystalline basement	Metasedimentary rocks with granitic intrusions.	None

SOURCE: U.S. Geological Survey Professional Paper 494-A, pp. A9-A11.

stable shelf of the continent east of them. The stable shelf at Grand Canyon lasted through late Precambrian, Paleozoic, and later time. It extends under all central United States, and the Precambrian rocks underlying it are widely exposed in the Canadian shield surrounding Hudson Bay.

The shelf and the shield form the stable nucleus of North America. Death Valley has been part of the unstable, mobile edge of the continent. The contrast in geology between Death Valley and Grand Canyon illustrates the broader structure of the continent. At Grand Canyon the formations are thin and nearly horizontal; at Death Valley they are thick and greatly deformed by folding and faulting and by igneous activity that melted some rocks and led to intrusions of granite and to volcanism.

Later Precambrian and Paleozoic and Triassic rocks began as marine sedimentary deposits below sea level. They were deposited in the ancient seas that flooded the geosynclines along the mobile belt at the western edge of the stable part of the continent. Subsequently they were uplifted, folded, faulted, and altered by igneous activity (chap. 6).

During the Jurassic and Cretaceous periods Death Valley was sharply elevated and the deformed rocks were eroded. In early Tertiary time igneous activity, including volcanism and granitic intrusions, reached a maximum, and the eruptive rocks and the sediments derived from them collected in structural basins similar to but located differently from the present ones. The later Cenozoic formations are mostly playa and fan gravel deposits eroded from the mountains and washed into the basins as we know them.

KINDS OF ROCKS AND THEIR LOCATIONS

The geologic map of Death Valley (fig. 65) shows the distribution of major groups of rock formations. The pages that follow describe the rocks more fully and indicate some of the easily accessible locations where they may be seen.

Precambrian Rocks   Precambrian rocks (table 4) representing the crystalline basement may be seen in the steep front of the Black Mountains and in some places close to the highway near Badwater and farther south. Some smaller outcrops are at the head of Galena Canyon and along the east front of the Panamint Range north of Hanaupah Canyon. In the Panamint Range these

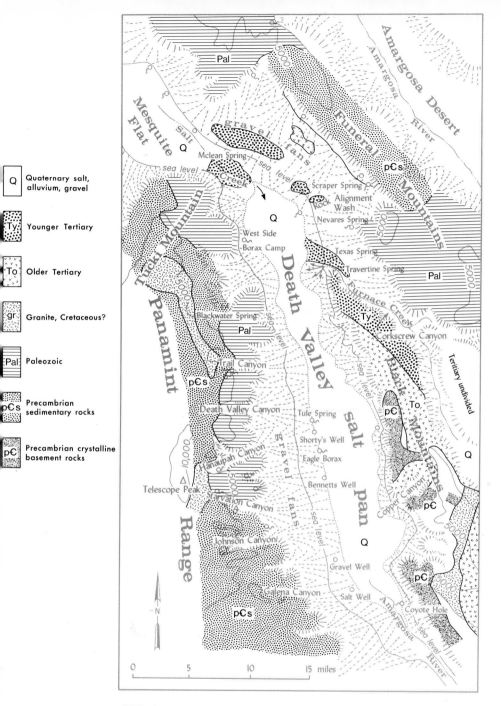

Quaternary salt,
Q alluvium, gravel

Ty Younger Tertiary

To Older Tertiary

gr Granite, Cretaceous?

Pal Paleozoic

pCs Precambrian
sedimentary rocks

pC Precambrian crystalline
basement rocks

FIG. 65. Geologic map of Death Valley.

old rocks include schist (fig. 66) and gneiss which probably were once sedimentary rocks, now greatly altered by metamorphism. A coarse facies of the gneiss is characterized by large feldspar crystals known as augen (fig. 67). The feldspar augen are 0.25 to 1 inch long and constitute 10 to 40 percent of the rock. The matrix is white quartz and black coarse-grained mica (biotite).

Overlying the crystalline basement is the Pahrump Series, consisting of sedimentary rocks only slightly metamorphosed. In the southern part of the Panamint Range the Pahrump is represented by the Crystal Spring Formation (fig. 68), about 2,000 feet thick. At the base is cemented gravel (conglomerate) which grades upward to sandstone, now metamorphosed to quartzite, which in turn grades upward to cemented mudstone (shale). The upper part of the formation is dolomite, a rock related to limestone but composed of magnesium carbonate in addition to calcium carbonate. The carbonate rocks are intruded by sheets of black igneous rock (diabase) and at the contacts the carbonates are altered to talc, which is extensively mined. The Death Valley region is one of the nation's important sources for talc. Elsewhere in the region two other formations are included in the Pahrump Series (see table 4).

The next three Precambrian formations overlying the Crystal Spring Formation in the southern Panamint Range are, in order, Noonday Dolomite, Johnnie Formation, and Stirling Quartzite. The Noonday consists of 800 feet of dolomite. The Johnnie Formation (fig. 69), more than 4,000 feet thick, has a lower member of sandstone or quartzite interbedded with dolomite and is gradational with the Noonday. A middle member consists of interbedded siltstone and dolomite, and the upper part of the formation is shale and quartzite. Uppermost of the Precambrian formations is the Stirling Quartzite (fig. 70), 2,000 feet thick.

No fossils have been found in these rocks or in their equivalents in other regions, yet a complex fauna suddenly appears at the beginning of the Paleozoic. Assuredly there must have been animals without hard parts in the seas represented by the

FIG. 66. Platy Precambrian schist in South Fork of Galena Canyon. Plates, almost vertical, trend northwest.

FIG. 67. Coarse-grained Precambrian rock known as augen gneiss contains large crystals of feldspar (white) and of biotite. Although the rock is about a billion years old, the radiometric age of the biotite crystals is only 12 to 14 million years (middle Tertiary), because biotite was formed or altered (metamorphosed) during Miocene volcanic activity. Photographed at foot of spur next north of Hanaupah Canyon.

Precambrian formations. Animals without shells or skeletons are rarely preserved as fossils; only the hard parts, not the fleshy tissues, are preserved. When, and in what form, did life first appear on earth? If the earth formed by condensation and cooling of a dust cloud, life could not exist until the ground had cooled sufficiently to permit the fall of rain and to permit the water to run off the land and collect in seas. Until the first rains there was no soil, and almost certainly there was no life on earth. If the theories that the earth's crust formed about 4.6 billion years ago and that the first rains and runoff occurred nearly 4 billion years ago are correct, the oldest Precambrian rocks in Death Valley

FIG. 68. View of Crystal Spring Formation northwest across South Fork of Galena Canyon. Light-colored beds at base of hill are quartzite (qtzte), which grades upward into shaly dolomite (do). Above dolomite is a dark, intrusive sheet of diabase (di); above and below diabase, where it is in contact with dolomite, are light-colored talc-bearing beds (t). Dark rock on top of hill is massive dolomite; light-colored formation above it is Noonday Dolomite (N), with still higher ledges of Johnnie Formation (J).

FIG. 69. Late Precambrian Johnnie Formation on north side of Six Spring Canyon, a mile above canyon mouth. Hilltop (center) is capped by quartzite included in upper part of Johnnie Formation. Below quartzite is a dark cliff of purple shale member. Farther down are well-stratified light-colored beds of dolomite and upper part of yellow shale member.

(and in Grand Canyon) were formed after the first soils and sediments had collected and probably after the first organisms had developed.

Older Paleozoic Rocks   About 8,500 feet of Cambrian rocks, the earliest of the Paleozoic, are present in the Death Valley region (table 4). The Wood Canyon Formation, about 2,500 feet thick, is at the base of the Cambrian. It overlies and is gradational with the Stirling Quartzite, suggesting that sedimentation was continuous during the time the two formations were deposited. The best exposures, along Blackwater Canyon (fig. 71), may be reached by

FIG. 70. Detail of bedding in late Precambrian Stirling Quartzite at mouth of Johnson Canyon.

FIG. 71. Cambrian formations along divide between Tucki and Blackwater washes. In foreground, extending as far as white bed of Zabriskie Quartzite (€z), is Wood Canyon Formation. The oldest recognizable fossils in Paleozoic formations in Death Valley region are fragments of trilobites which occur in shale beds (sh) crossing saddle in foreground. Above Zabriskie Quartzite is highly fossiliferous Carrara Formation (€c); dark dolomite capping most distant hill is lower part of Bonanza King Formation (€b).

a jeep trail; the direct route crossing the valley below Furnace Creek fan has been closed, even for four-wheel-drive vehicles, by the Park Service. The most accessible outcrops are in Echo Canyon and at Death Valley Buttes. The oldest rocks in the Death Valley region containing fossil animal remains are in the Wood Canyon Formation.

The fossils are not very exciting to look at, for they are mostly fragments of trilobites, smaller and more fragmental than those illustrated in figure 72. Yet they record one of the most amazing and least understood episodes in all geologic history. How might one explain the sudden appearance of animals with hard parts at the beginning of the Cambrian, in the midst of what apparently was a continuous sequence of sediments? There must have been a long history of prior development. The sudden (geologically speaking) appearance of fossils at the beginning of the Cambrian probably was not a population explosion. More likely, oceanic water chemistry was passing a critical threshold so that dissolved minerals could be used by the organisms. For the present, however, these beds with their fossils in Death Valley remain a mystery.

Next above the Wood Canyon Formation is the Zabriskie Quartzite from a hundred to several hundred feet thick. The Zabriskie can be a tough rock that forms ledges, but usually it has been so crushed that it is only a mass of granulated quartz fragments. No fossils were found in the formation.

Overlying the Zabriskie is the Carrara Formation, consisting of about 1,000 feet of richly fossiliferous beds that are transitional between the underlying formations of sandstone and shale and the overlying formations, which are largely limestone and dolomite. The formation consists of alternating beds of shale, silt, and limestone. Some beds are composed so largely of fragments of trilobites that they are referred to as trilobite trash beds (fig. 72). There are numerous exposures on both sides of Echo Canyon, just above the canyon mouth, and in the upper part of Trail Canyon.

Uppermost of the Cambrian formations are the Bonanza King (fig. 73) and the Nopah formations, which are largely dolomite.

FIG. 72. Trilobite trash bed, Carrara Formation. Abundant trilobites are characteristic of Cambrian formations. Total width of specimens, 7 inches.

FIG. 73. Bonanza King Formation, as seen in this view in upper Echo Canyon, is mostly dark-colored, massive, cavernous dolomite.

The most accessible outcrops are in Trail and Echo canyons. The Bonanza King Formation, about 3,000 feet thick, consists largely of massive cavernous dolomite, but near the middle of the formation are two light tan sandy and shaly zones about 200 feet apart, each about 50 feet thick. The upper one contains ovaloid brachiopods (fig. 74) and trilobites. The several hundred feet of dark dolomite above and below these fossiliferous beds have no fossils.

The Nopah Formation (fig. 75), marking the top of the Cambrian, is 1,200 to 1,500 feet thick. At the base are fossiliferous shaly beds. or trash beds, 100 feet thick containing abundant fragments of brachiopods and trilobites. The rest of the formation, consisting of banded light and dark gray dolomite, is not fossiliferous.

Ordovician formations, which overlie the Cambrian, are well exposed in readily accessible outcrops along the road in Trail Canyon (fig. 76). They include the Pogonip Formation at the

FIG. 74. Ovaloid brachiopods in light tan, shaly, sandy beds near middle of Bonanza King Formation. With brachiopods are fragments of trilobites. Total width of specimens, 7 inches.

FIG. 75. Uppermost of Cambrian formations is Nopah Formation, chiefly dolomite banded dark and light, as seen in this view in Trail Canyon. The dolomite contains chert nodules; it is not fossiliferous. Shaly beds at base of formation contain fragments of trilobites and brachiopods.

base, the light-colored Eureka Quartzite, and, at the top, the dark Ely Springs Dolomite. Total thickness is about 2,500 feet.

The Pogonip Formation in the Death Valley area has a basal unit of dolomite and a middle unit of reddish limy shale and sandstone, including trash beds containing fragments of fossils. The top of the Pogonip is a dolomite member which forms cliffs and locally contains large gastropods (fig. 77).

FIG. 76. Cambrian and Ordovician formations are well exposed along north side of Trail Canyon. Єn, Nopah Formation; Opl, lower dolomite member of Pogonip Formation; Opm, middle shale member of Pogonip; Opu, upper dolomitic member of Pogonip; Oe, Eureka Quartzite; Oes, Ely Springs Dolomite; Tv, Tertiary volcanic rocks. The volcanic rocks dip much less steeply than the Ely Springs Dolomite and rest unconformably on the dolomite.

FIG. 77. Abundant gastropods characterize upper cliff-forming dolomitic member of Pogonip Formation. Ruler is 6 inches long.

The Eureka, a massive vitreous quartzite about 350 feet thick (fig. 78), resembles the Zabriskie Quartzite. In many places it is granulated as a result of shearing during folding and faulting. Uppermost of the Ordovician formations, the Ely Springs Dolomite, 400 to 800 feet thick, is dark, thick-bedded dolomite (fig. 78) containing numerous fossils.

The Hidden Valley Dolomite of Silurian and Devonian age overlies the Ordovician formations (fig. 79). Many of its beds contain corals (especially genus *Halysites*) and crinoid stems up to half an inch in diameter. In thickness the Hidden Valley Dolomite ranges from 300 to more than 1,000 feet.

Above the Hidden Valley Formation is the Lost Burro Formation of Devonian age. About 2,000 feet thick, it is striped with alternate light and dark beds. Many of the dark beds are mottled

FIG. 78. Steeply dipping, light-colored Eureka Quartzite overlain by dark Ely Springs Dolomite at north foot of Tucki Mountain (mouth of Little Bridge Canyon).

with whitish streaks resembling chopped spaghetti. Common fossils are brachiopods and corals (fig. 80). The most accessible exposures of the Hidden Valley and Lost Burro formations are along the north foot of Tucki Mountain east of the mouth of Little Bridge Canyon.

FIG. 79. View northeast across east end of Tucki Mountain, Cottonball Basin, and Funeral Mountains. Formations that can be seen in Tucki Mountain range in age from Cambrian to Permian. €n, Nopah Formation; Op, Pogonip Formation; Oe, Eureka Quartzite; Oes, Ely Springs Dolomite; DSh, Hidden Valley Dolomite; Dl, Lost Burro Formation; M-P, Mississippian, Pennsylvanian, and Permian formations. (Photo by John H. Maxson.)

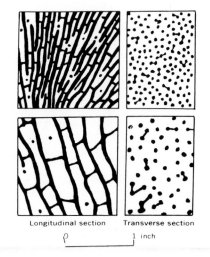

Longitudinal section   Transverse section

0   1 inch

FIG. 80. Diagrammatic sections of Devonian (upper) and Mississippian (lower) syringoporoid corals. In Devonian corals the corallites and connecting tubes are closely spaced; in Mississippian corals they are widely spaced. (From U.S. Geol. Survey Prof. Paper 494-A.)

Younger Paleozoic Rocks   Younger Paleozoic rocks include those of the Mississippian, Pennsylvanian, and Permian periods (roughly 375 to 225 million years ago). The formations in Death Valley, entirely of marine origin, record the continuation of the geosynclinal trough that extended northward across what is now Nevada to western Montana and the Canadian Rockies.

On Tucki Mountain, 1,700 feet of limestone of Mississippian age overlies the Lost Burro Formation. It includes the Tin Mountain Limestone and an upper member that still is not named, although sometimes correlated with the Perdido Formation farther north. The formations contain a considerable fauna of brachiopods, gastropods, corals, bryozoans, and crinoids. Above the limestone is the Rest Spring Shale. About 750 feet of it is exposed but the formation may be thicker than that because the shale is crushed between nearly vertical resistant formations. The formation is characterized by rounded nodules of dark chert, which look like golf balls. Some of the beds are ripple marked. The fossils include brachiopods, bryozoans, and fish scales. These outcrops are most easily reached from the road to West Side Borax Camp, sometimes referred to as Shoveltown, which runs along the salt pan at the east foot of Tucki Mountain.

The youngest Paleozoic formations in the Death Valley area, of Pennsylvanian and Permian age, record a major episode of late Paleozoic earth movements. The formations are well exposed at the east foot of Tucki Mountain, thrust faulted onto a quartzite thought to be the Stirling (see chap. 6) and crushed against the Rest Spring Shale. The formations include 2,500 feet of thin-bedded limestone, largely if not wholly of Pennsylvanian age, and limestone conglomerate 3,000 feet thick which is largely if not wholly of Permian age (fig. 81).

FIG. 81. Conglomerate of Permian age at east foot of Tucki Mountain. Cobbles and boulders in conglomerate are from Mississippian and Pennsylvanian formations, indicating an episode of folding or other deformation in late Pennsylvanian or Permian time.

The conglomerate records the Paleozoic deformation. It consists of well-rounded cobbles and small boulders as much as a foot in diameter, mostly limestone and dolomite but including some quartzite. They are fossiliferous and range in age from early Mississippian to late Pennsylvanian. Evidently the conglomerate began forming in very late Pennsylvanian or early Permian time as a result of a deformation that caused erosion of Mississippian and Pennsylvanian rocks.

Mesozoic Rocks   Rocks of Triassic age (ca. 225 to 180 million years ago) are exposed in Warm Spring Canyon and Butte Valley, near the southern end of the Panamint Range. The Triassic rocks are in part fossiliferous sediments and in part volcanic. Whereas the trough in which the Paleozoic formations were deposited trended northward across Nevada, the Triassic trough trended northwestward through what is now the Sierra Nevada. The big mass of granite forming the Sierra, called a batholith, developed in that trough in Jurassic and Cretaceous time. Melting of the Triassic rocks probably contributed to formation of the granite. The Triassic formations are the youngest marine deposits in Death Valley.

Rocks of Jurassic and Cretaceous age are not known in Death Valley. During those periods (180 to 60 million years ago), while the Sierra Nevada granite was being formed, the Death Valley area apparently was mountainous and was being eroded. Death Valley is at the eastern edge of a belt of isolated granitic intrusions rather like the clustered ones that form the Sierra Nevada. The isolated intrusions are younger than the Sierra Nevada granite, but they probably are satellites related to it. Three granitic intrusions in the Death Valley area are at Skidoo, Hanaupah Canyon, and the east foot of the Panamint Range north of Hanaupah Canyon. The intrusions spread laterally along the thrust faults; that is, they are floored intrusions (laccoliths). The igneous activity progressed to the surface and gave rise to volcanic eruptions in Tertiary time. A chaotic assemblage of igneous, metamorphic, and sedimentary rocks found in places along the thrust faults is referred to as the Amargosa Chaos (see chap. 6). The igneous activity may have started in late Cretaceous time, but continued into the Tertiary.

Tertiary Rocks   At the south end of the Black Mountains, in the vicinity of Virgin Spring and Jubilee Pass, Tertiary rocks (60 to 2 or 3 million years ago) include conglomerate, shale, freshwater limestone, lavas of various kinds, and volcanic ash. These rocks, originally part of an orderly sequence like the one at the north end of the mountains, are now in masses of irregular shape and size (up to 1,000 feet long) in a crushed mass (breccia) forming the Amargosa Chaos. With the Tertiary rocks are blocks of granite and of Precambrian sedimentary rocks. The shingled mass of individual blocks has been pushed (thrust faulted) onto Precambrian rocks.

Cutting the Precambrian rocks in the Black Mountains above Badwater are vertical sheets (dikes) of various kinds of igneous rocks. These are sheared off at the thrust fault under the overlying Chaos, but evidently the dikes were once the source of considerable volcanic activity.

Tertiary rocks at the north end of the Black Mountains, aggregating 13,000 feet in thickness, are more orderly. They overlap and are faulted against the Precambrian core of the Black Mountains, as can be seen from the highway just north of Badwater. The Tertiary formations consist of volcanic rocks and sedimentary rocks; the percentage of the latter increases northward away from the dike swarm cutting the Precambrian rocks above Badwater.

These Tertiary rocks are divided into an older set of beds, the Artist Drive Formation (fig. 82), probably Oligocene to early Pliocene in age, and a younger set, the Furnace Creek Formation (fig, 83) of Pliocene age. The Tertiary rocks, mostly shale, siltstone, and sandstone, are probably playa deposits. There are

FIG. 82. Volcanic rocks and derived sediments form front of Black Mountains at Artists Drive.

FIG. 83. Panorama of Furnace Creek Formation in Gower Gulch as seen from Zabriskie Point.

also conglomerates as much as 100 feet thick containing cobbles of granite and late Paleozoic formations as well as volcanic rocks.

Fossils are scarce in these rocks. The uppermost playa beds in the Artist Drive Formation have yielded diatoms, microscopic plants indicative of early Pliocene age. Volcanic rocks on the west side of Death Valley which are like those in the Artist Drive Formation, have been dated radiometrically as late Miocene.

The Pliocene Furnace Creek Formation, which overlies the Artist Drive Formation at the north end of the Black Mountains, extends northward from there to the fault along the southwest side of the Miocene(?) formations in the Kit Fox Hills* (fig. 84). It probably underlies Cottonball Basin because it reappears northwest of the basin in the uplift at Salt Creek Hills. The formation is more than 5,000 feet thick and consists largely of fine-grained playa deposits containing borate minerals. Diatoms from the playa beds indicate that the formation spans most of Pliocene time. The playa extended northwestward along Furnace Creek Wash and across Cottonball Basin to Salt Creek Hills; it trended 45 degrees to the present Death Valley.

*According to Wesley Hildreth of the Park Service (personal communication, 1970), rounded cobbles of welded tuff, dated as Upper Miocene in the Grapevine Mountains and at the Nevada test site, have been found in the Tertiary of the Kit Fox Hills. They might suggest a Pliocene age for those rocks.

A conglomerate 200 feet thick at the base of the Furnace Creek Formation, well exposed and readily accessible at the front of the Black Mountains near Gower Gulch, contains cobbles and boulders of Paleozoic limestone and dolomite measuring up to a foot in diameter. Among them are fossiliferous limestone from the Pennsylvanian and Permian formations, granular limestone with crinoid fragments that almost certainly are from the Mississippian Tin Mountain Limestone, and limestone with brachiopods from the Devonian Lost Burro Formation. About 65 percent of the cobbles are Paleozoic carbonate rocks, 10 percent are quartzite like the Zabriskie or Eureka quartzites, 20 percent are volcanic rock, and 5 percent are granite or other miscellaneous types. The granite does not look like the granites at Skidoo and Hanaupah Canyon. Either these rocks were derived from the northwest across what is now Death Valley, or a mountain of granite and late Paleozoic rocks has been faulted downward under the younger deposits.

Whereas the conglomerate at the base of the Furnace Creek Formation seems to have been derived from the northwest, the younger and higher conglomerate that outcrops at the mouth of Furnace Creek (shown at the right in figure 83) seems to have been derived mostly from the Black and Funeral mountains. Cobbles in this conglomerate look as though they came from early Paleozoic formations, probably those in the southern part of the Funeral Mountains.

On the basis of contrasts in the sources of sediments in the Furnace Creek Formation, one can attempt to visualize the geography at the time the formation was deposited, some 5 to 10 million years ago.

Probably there was a playa elongated northwestward at the present site of the north end of the Black Mountains, along the north base of a volcanic pile (Artist Drive Formation) which also extended northwestward across what is now the Middle Basin of Death Valley. This volcanic pile included the volcanic rocks now above Badwater and those north of Trail Canyon on the opposite

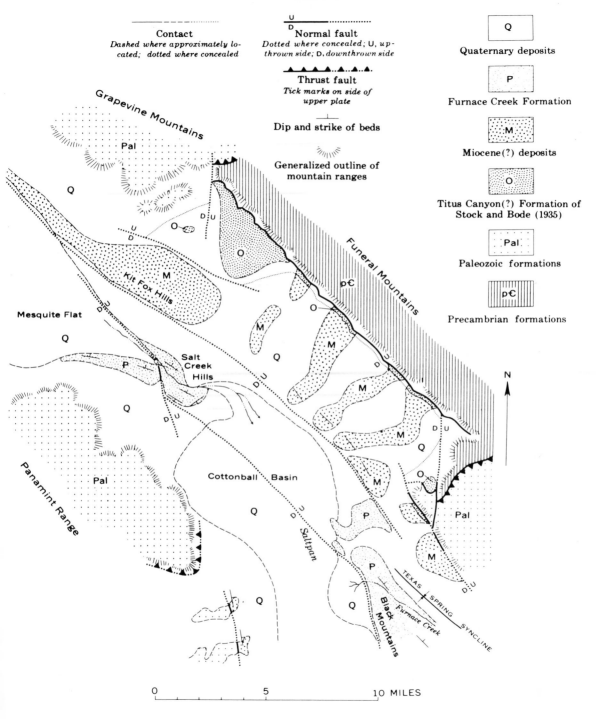

EXPLANATION

Contact
Dashed where approximately located; dotted where concealed

Normal fault
Dotted where concealed; U, upthrown side; D, downthrown side

Thrust fault
Tick marks on side of upper plate

Dip and strike of beds

Generalized outline of mountain ranges

Q — Quaternary deposits

P — Furnace Creek Formation

M — Miocene(?) deposits

O — Titus Canyon(?) Formation of Stock and Bode (1935)

Pal — Paleozoic formations

pЄ — Precambrian formations

Grapevine Mountains

Funeral Mountains

Panamint Range

Black Mountains

Kit Fox Hills

Mesquite Flat

Salt Creek Hills

Cottonball Basin

Saltpan

Furnace Creek

TEXAS SPRING SYNCLINE

N

0   5   10 MILES

FIG. 84. Sketch map of Tertiary formations around Cottonball Basin. (From U.S. Geol. Survey Prof. Paper 494-A.)

side of the present valley. Most of the fine-grained sediments deposited in the playa were derived from the volcanic pile, but deposition was interrupted by influxes of coarse gravels from the northwest, probably in response to faulting in that area. During the second half of Pliocene time an increasing amount of sediment was brought from the south end of the Funeral Mountains, probably in response to uplift there.

Distribution of salts in the upper playa beds suggests that the central part of the Pliocene playa was a short distance east of the present edge of the salt pan in Cottonball Basin; the playa beds there are highly saline, as can be seen in the hills at Mustard Canyon. At the East Coleman Hills the playa beds are at least 2,400 feet thick and contain sulfates and borates, which suggests a position nearer the edge of the Pliocene playa (see chap. 3). About a mile farther southeast the upper playa beds are 1,300 feet thick and contain abundant veins of gypsum. Still farther southeast, along the flank of the syncline at Texas Spring, these beds continue to thin and contain less sulfate, almost no chlorides, and more carbonate and granular volcanic ash.

The lower playa beds of the formation contain thick deposits of gypsum and of borates that were produced while the 20-mule teams were operating. These deposits are southeast of those in the upper playa beds, as if the sulfate-borate zone had shifted northwestward toward Cottonball Basin. Today that zone is at the edge of the salt pan.

Other Tertiary formations exposed around the north end of Cottonball Basin are less well known. The older of these deposits consist of shale, sandstone, and conglomerate and look very much like the formations in Titus Canyon in the Grapevine Mountains, where Oligocene vertebrate fossils have been found. The beds, at least 1,500 feet thick, are faulted against the Precambrian rocks of the Funeral Mountains and in part are derived from those mountains. They were deposited against the fault at the front of the mountains and subsequently were displaced by renewed movement along the fault.

One of the curious and seemingly characteristic features of these rocks is the habit of the conglomerates. The cobbles are aligned parallel to the bedding but are broken by transverse fractures at right angles to the bedding (fig. 85).

Tertiary rocks forming the Kit Fox Hills, and those in the fault blocks southeastward from there to the foot of the Funeral Mountains, are regarded as of Miocene age although the only

FIG. 85. Detail of Tertiary gravel bed in Titus Canyon Formation (?) exposed near Keane Spring. Cobbles are aligned with bedding which dips steeply to right. Flat fractures at right angles to bedding dip gently to left and extend through cobbles, which are displaced as much as 0.25 inch. These fractures have healed, and the fractured cobbles can be recovered intact.

known fossils are nondiagnostic plant stems and animal tracks. The rocks constitute an intermediate belt between Oligocene and Pliocene rocks. They are estimated to be 4,000 feet thick. Pebbles in the conglomerates suggest a source in the northern part of the Funeral Mountains.

The extensive Tertiary volcanic rocks west of Death Valley and overlapping the west flank of the Panamint Range are described in chapter 6.

Quaternary Rocks   Quaternary rocks (the past 2 or 3 million years) include the playa deposits forming the flat floor of Death Valley, with its salt crust, and the gravels in the fans that rise from the edge of the playa to the bordering mountains. As these areas are strikingly different in environment, they are treated in separate chapters (2, 3, 4). In this section only the earliest of the Quaternary fan gravel deposits are discussed; they are old enough to have been so faulted, folded, and eroded that they no longer retain the fan form.

The old fan gravels are cemented and are referred to as fanglomerate. The Funeral Formation outcrops extensively in fault blocks that extend northwest along Furnace Creek Wash and along the east side of Cottonball Basin. These gravels filled the southeast end of the Pliocene playa recorded by the Furnace Creek Formation. Although the Funeral Formation overlies the light-colored playa beds of the Furnace Creek Formation, both are faulted and steeply tilted and both are partly buried by younger fan gravels that washed across them (fig. 86).

The Funeral Formation also outcrops in the Artist Drive fault blocks where it unconformably overlies the volcanic and other rocks of the Artist Drive Formation, which had been faulted before the Funeral Formation was deposited. The Funeral Formation there includes flows of basaltic lava beds of volcanic ash up to 4 feet thick. The volcanic activity in that area continued to Quaternary time.

Other outcrops of the Funeral Formation are at Mormon Point

and in extensive areas along the east side of Emigrant Wash. Along Emigrant Wash the formation is at least 3,000 feet thick. In both areas the fanglomerate was deposited against a faulted surface of Precambrian rocks and in both areas it has been displaced by renewed movement along those faults. The Funeral Formation also overlies and is faulted against the west flank of the Skidoo granite.

The gravels contained in the conglomeratic layers of the Funeral Formation are rocks derived from bordering mountains. The formtion was deposited while the mountains were being elevated to their present position and became faulted, tilted, and eroded; the fan form was destroyed as the deformation progressed. In the Park Village area, for example, the faulted fanglomerate forms a ridge about 350 feet high which trends north-

FIG. 86. Steeply tilted Upper Pliocene playa deposits in Furnace Creek Formation (Tpf) and Lower Pliocene gravel deposits in Funeral Fanglomerate (Qf). The cemented fanglomerate is cut by banded veins of calcite (thin white streaks) known as Mexican onyx. View is southeast along escarpment crossed by Cow Creek southeast of Park Service service area.

ward roughly along the contour of the gravel fans that rise eastward to the Funeral Mountains. The drainage now is incised across the ridge, and younger gravels lie on both sides of it. Similarly at Salt Creek Hills the Funeral Formation is raised in a structural dome and there Salt Creek is incised into it.

In both these areas the Funeral Formation is displaced less than the underlying Furnace Creek Formation; clearly, therefore, the faulting that occurred during Pliocene time continued. It is likely that the deformation progressed in small increments over a long period of time, so that the streams were able to cut their way through the ancient playa beds and fanglomerate as they were uplifted, a process known as antecedence and first described by John Wesley Powell in his report on his trip down the Colorado River in 1869-70.

6   How the Rocks Broke

Death Valley is part of the southwestern desert of basins and
ranges lying south and west of the Colorado Plateau and extending westward to the Sierra Nevada. The basins and ranges east of
the Colorado River, in southern Arizona, have well-established
drainage systems such as the Gila and Bill Williams rivers. The
Mojave Desert west of the Colorado River has no such streams;
the drainage disintegrates in a cluster of irregular basins and
ranges. Death Valley is at the south end of another cluster of
closed basins, an area referred to as the Great Basin. Great Salt
Lake is in the northeast corner of the Great Basin; Carson Sink
and the lakes north of Reno are in the northwest part.

   The valleys in this part of the country are structural valleys, not
caused by erosion. They have been faulted down and the mountains have been raised (fig. 87). In general, the faults lie along
the eastern sides of the valleys, the western sides of the mountains, and the fault blocks have been tilted to the east. As a
consequence, in most valleys the bordering mountains have
precipitous escarpments facing the valley along the east side and
long, less steep slopes rising westward.

   High on the mountains in the bedrock formations of limestone
and shale may be found fossil seashells. The limestone and shale

were deposited in the ancient sea where the marine animals lived, and subsequent earth movements, folding and faulting, have raised the rocks to their present high elevations. The sea was never that high. This phenomenon illustrates a fundamental principle of geology about the relation of a landscape to the rocks that form it.

A mountain or a valley records chapters of earth history which are widely separated in time and only slightly related. The older, more obscure chapter is concerned with the history of the rocks forming the particular mountain or valley; the younger chapter is concerned with how the mountain or the valley developed the topographic form it now has. The Paleozoic rocks in Death Valley (see chap. 5) are 225 to 600 million years old; the Precambrian rocks are even older. But the Panamint Range, the Black Mountains, and Death Valley, in the form they have today, developed largely during the past 4 or 5 million years. The oldest gravels at the surface of the fans probably are no older than 500,000 years, and the salt crust on the salt pan is less than 2,000 years old.

The vast stretches of geologic time defy the imagination and, like the public debt, are difficult to comprehend. Benjamin Franklin, speaking of debts, said that a "small leak can sink a big ship"; in geology small events repeated over and over again in geologic history can cause major changes. A foot of uplift in 500 years is not an exciting rate of earth movement; indeed, parts of Death Valley are moving today at livelier rates than that. But such a movement, continued for 5 million years, can build a mountain range as high as the Panamint. The mountains and the valleys were not broken by huge convulsions, but rather by small increments of movement over a very long period of time.

PRESENT-DAY EARTH MOVEMENTS

Geologic evidence (chap. 3) indicates that Death Valley has been tilted 20 feet to the east since the valley was flooded by the

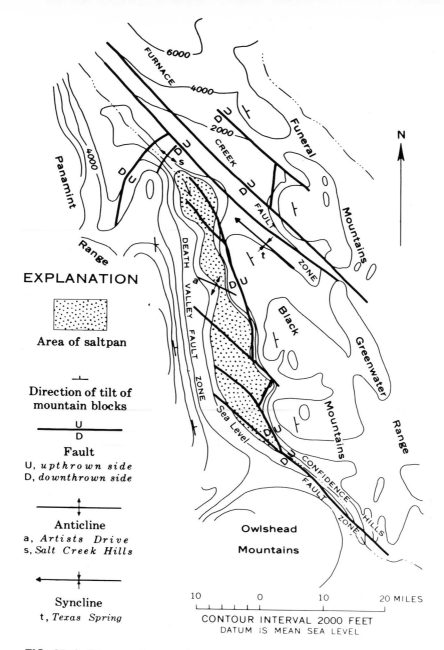

EXPLANATION

Area of saltpan

Direction of tilt of
mountain blocks

U
D
Fault
U, *upthrown side*
D, *downthrown side*

Anticline
a, *Artists Drive*
s, *Salt Creek Hills*

Syncline
t, *Texas Spring*

10   0   10   20 MILES

CONTOUR INTERVAL 2000 FEET
DATUM IS MEAN SEA LEVEL

FIG. 87. In Pliocene time, perhaps as recently as 5 million years ago, the Death Valley region consisted of a series of northwest-trending mountains and structural valleys, mostly outlined by the principal faults of the region and the anticlines as shown here. The faulting and folding began far back in Tertiary time, but on these faults and folds there has been displacement during the Quaternary, some of it as recently as Holocene time. The mountains presently are being tilted in directions indicated by symbols. The north-trending valley and mountains are largely the result of the very late Tertiary and later structural movements. (From U.S. Geol. Survey Prof. Paper 494-A.)

Holocene lake that produced the salt pan; the eastern shore of
the lake is that much lower than the western shore. That the
valley is still moving seems indicated by warps in the floodplain
which now isolate some parts of it. Middle Basin, for example, is
separated from Badwater Basin by a warp of the floodplain which
is 18 inches high; the northeastern part of Cottonball Basin is
isolated from the central part by a warp about 6 inches high. Two
other warps of the floodplain are expressed by sweeping curves
in the drainage that lies between the foot of the Black Mountains
and the eastern edge of the rock salt at Mormon Point and north
of Copper Canyon (fig. 88). Both warps, as well as the one crossing the floodplain at the south end of Middle Basin, represent
renewed movement along old structures.

On the basis of evidence indicating crustal instability, the
United States Geological Survey installed tiltmeters at several
stations in the Panamint Range, Death Valley, and the Black
Mountains, to measure tilt that may be continuing. The installa-
tions are simple enough in principle (fig. 89). Three concrete
posts anchored on bedrock and capped by metal plates which are
approximately level are installed to form a triangle with sides
about 200 feet long. Two pots, built to fit securely on the metal
caps, are half filled with water and linked by two hoses, one connecting the air chambers and the other connecting the water. The
pots are then sealed. A window on the side of each pot, which
shows the water level, is equipped with a micrometer screw
having a needle that can be moved to touch the water surface; the
reading is made when the water surface first responds to the
needle point. Differences in level of the posts can be measured by
traverses around the triangle to control the error. Tilt amounting
to about 6 microns (.006 mm) in 200 feet can be measured.
Observations over a three-year period (1958-1961) showed that
the structural blocks comprising Death Valley and the bordering
mountains are being measurably tilted. The record reveals that
the tilting on a particular fault block did not progress consis-
tently in one direction, although in all instances tilting was in the

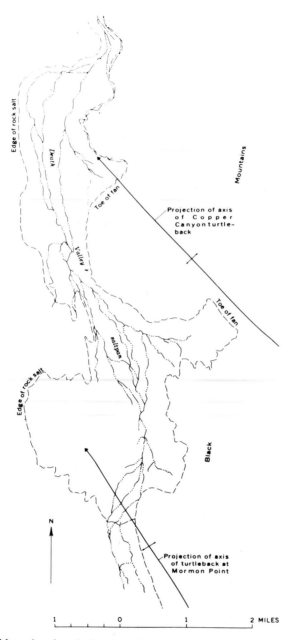

FIG. 88. Map showing drainage in Death Valley salt pan deflected at projected axes of turtlebacks (anticlinal ridges) in Black Mountains. (From U.S. Geol. Survey Prof. Paper 494-A.)

direction indicated by the geology. Seemingly a block would tilt at an excessive rate, and then the tilt would partly reverse itself as if the blocks were being shaken back into place by tremors. It would be exciting to discover the nature of longer term trends, but unfortunately the Geological Survey has discontinued the measurements.

Continued earth movement is suggested by changes in the level of bench marks in Death Valley (fig. 90), as shown by leveling surveys in 1907 and resurveys in 1933 and 1942-43. At all eight stations reexamined in 1942-43, the level averaged 3 to 4 inches

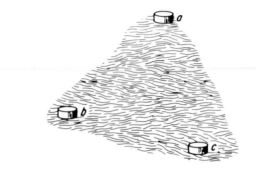

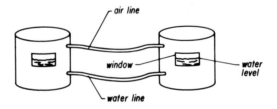

FIG. 89. Principle of tiltmeters used in Death Valley to measure present-day earth movement. Upper diagram: Tiltmeter station consists of three concrete posts (*a, b, c*) forming equilateral triangle, each leg measuring about 180 feet. Each post is fitted with a metal cap. Lower diagram: Two sealed pots half filled with water and closely fitted to metal caps are placed on two of the posts; the air chambers are connected by a hose, and a second hose connects the water in lower half of the pots. A micrometer screw inside each pot measures height of the water and the three posts that form triangle are checked to determine height of each. Successive readings over a period of time show whether triangular area has been tilted.

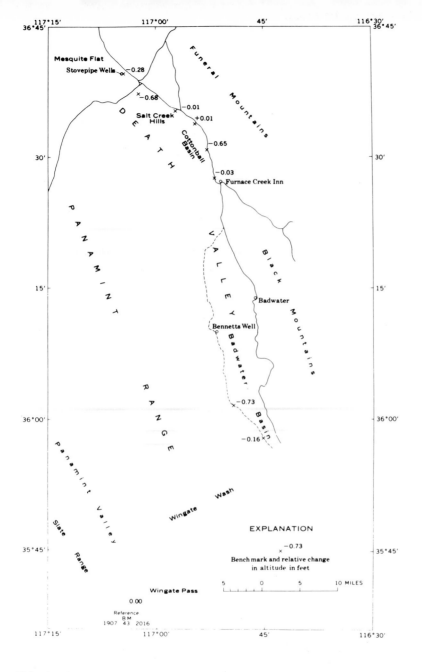

FIG. 90. Relative changes in altitude of bench marks in Death Valley between unadjusted 1907 U.S. Geological Survey data and adjusted 1933 and 1942-43 U.S. Coast and Geodetic Survey data. The changes are relative to bench mark BM 1907 43 2016 near Wingate Pass. (From U.S. Geol. Survey Prof. Paper 494-A.)

lower than in 1907. A loss of that much in thirty-five years amounts to roughly 1 inch in ten years or about 20 feet in 2,000 years, which is of the right order of magnitude. The evidence is difficult to use, however, because engineers attribute the differences to errors in surveying. Since I do not believe that engineers are all that incompetent, I ascribe the differences to downwarping of Death Valley. The displacements are about the amount to be expected and lie in the right directions. It is to be hoped that the Coast and Geodetic Survey and the Geological Survey can be persuaded to undertake additional precision surveys to measure possible, and probable, earth movements in Death Valley.

DEATH VALLEY'S LAST MAJOR EARTHQUAKE

When the last big earthquake in Death Valley occurred, about 2,000 years ago, the valley floor was tilted to the east and half of the 20-foot displacement of the shoreline of the Holocene lake was taken up in 10 feet of displacement along a fault at the foot of the Black Mountains (figs. 91, 92). This fault can be traced most of the length of the Black Mountains, from near Furnace Creek to Mormon Point. The faulting occurred after the desert varnish was formed but earlier than some of the Indian occupations. At the readily accessible fan just south of Furnace Creek fan, for example, Indians built mesquite storage pits in the rubble along the fault escarpment (see pit D in fig. 93).

What happened to Death Valley during the earthquake 2,000 years ago was duplicated almost exactly in the Hebgen Lake earthquake west of Yellowstone Park in 1959. The lake bed there was tilted 20 feet to the northeast, and half of the displacement was by faulting. Docks on the southwest shore of the lake were raised out of the water; roads and homes along the northeast shore were submerged. A building that extended across the fault was broken into two parts, as were roads that crossed the fault. It

FIG. 91. View south across marsh at Badwater. On left side of fan can be seen escarpments of last 10 to 20 feet of faulting along front of Black Mountains.

FIG. 92. Last big earthquake in Death Valley, about 2,000 years ago, accompanied faulting that produced this escarpment near road just south of Furnace Creek fan. The escarpment is buried where it crosses mouths of canyons issuing from Tertiary formations at north end of Black Mountains.

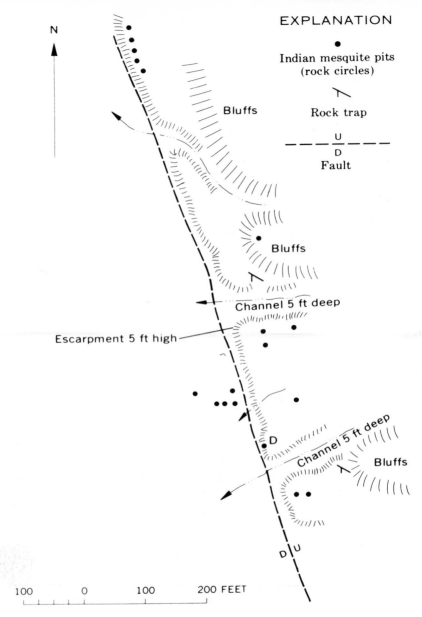

FIG. 93. Map of Indian sites along escarpment of a Holocene fault at foot of Black Mountains 3 miles south of Furnace Creek Wash. Fault and escarpment are older than Indian mesquite storage pit at *D*, which is built in colluvium overlapping the scarp. (From U.S. Geol. Survey Prof. Paper 494-A.)

all happened in a matter of minutes, and very likely the Death Valley earthquake had the same duration. One can only wonder about the reaction of Indians who may have been in Death Valley on that day.

The shaking of the ground at Hebgen Lake triggered a tremendous landslide into Madison Canyon, a slide big enough to dam the Madison River and create a new lake. In Death Valley only one landslide is known which might have been triggered by earth tremors, and it seems to be older than the last Death Valley earthquake. The slide is in the canyon at the front of the Black Mountains, 3 miles south of Badwater (fig. 94). As the lower end of this slide was displaced about 75 feet by a late Pleistocene fault, the slide must be old.

FIG. 94. Rock slide at front of Black Mountains 3 miles south of Badwater was probably triggered by an earthquake. Lower end of slide has been displaced 75 feet by late Pleistocene faulting.

EARLIER EARTHQUAKES

The Holocene faulting represents renewed displacement on faults
that had been active for a long time. Evidence of late Pleistocene
displacement, for example, can be seen along the foot of the
Black Mountains most of the way from Furnace Creek to Mormon
Point. At Furnace Creek, Pleistocene terrace gravels have been
uplifted about 75 feet (see fig. 56). At Mormon Point, both Holocene and Pleistocene stages of faulting can be seen (fig. 95);
south of Furnace Creek, in the vicinity of Desolation Canyon, the
Holocene fault scarp is exposed (fig. 92), and the older and
greater Pleistocene displacement is recorded by hanging valleys
(fig. 57). The hanging valleys are old ones having rounded cross
profiles. When their courses became disrupted by the faulting,
the streams cut narrow gorges in the bottoms of the valley seg-
ments that were uplifted. The contrasting valley shapes that
resulted from the faulting constitute what is known as a topographic unconformity.

Two stages of displacement can also be seen in the escarpment
of gravel west of Shorty's Well; the older gravel, no. 2 (chap. 4),
has been displaced about 50 feet along a fault on which renewed
movement has displaced no. 3 gravel by 6 feet. That the hanging
valleys are approximately the age of no. 2 gravel is suggested by
the fact that benches of the same gravel near the mouth of
Furnace Creek (fig. 56) are displaced by about the same amount
as nearby hanging valleys. Also, no. 2 gravel is domed and
faulted at the uplifts at Mustard Canyon Hills (fig. 96), East Coleman Hills, and Salt Creek Hills (fig. 14). No. 2 gravels are down-
folded into the syncline along Furnace Creek Wash. These,

FIG. 95. Two stages of faulting can be seen in gravels beside road at Mormon Point. Deep washes were incised into gravels, uplifted 75 feet by Pleistocene fault. Renewed Holocene activity on fault has displaced gravel in washes by about 10 feet.

FIG. 96. Late Pleistocene gravels (dark) domed across uplift at Mustard Canyon. Light-colored beds are Pliocene playa deposits of Furnace Creek Formation.

however, show less downfolding than those in the older Funeral Formation, and the latter in turn are less folded than those in the still older Furnace Creek Formation of Pliocene age, supplying further evidence that the deformation occurred in multiple stages over a long period of time.

There is some evidence that both the Black Mountains and the Panamint Range were tilted northward prior to the deposition of no. 2 gravel. Along the front of the Black Mountains the hanging valleys are lower toward the north than they are at the south, as if the south end of the mountains had been uplifted more than the north end. Northward along the east foot of the Panamint Range the fans extend progressively farther into the mountain valleys, a relationship that can be seen from the highways on the valley floor. At the south end of the mountains the spurs are nearly straight and the fans apex at the mouths of canyons draining the mountains. Farther north, as at Six Spring Canyon, the mountain front is frayed and the gravel extends into the canyon as a narrow deposit along the bottom. At the north end of the salt pan, at Blackwater and Tucki Wash, the fan embays the mountain front and buries much of the spurs (fig. 49). This part of the Panamint Range, which seems to have been tilted northward, is partly buried under its own debris.

VALLEYS BEFORE DEATH VALLEY

Death Valley, since its beginning, has received sediments eroded from the mountains around it. Gravity and other geophysical measurements indicate a total of 8,000 feet of fill in the valley. Despite the large amount of fill, the rate of subsidence of the valley has been faster than the rate of filling, and the floor is now below sea level. During the deposition of no. 2 gravel, when large boulders were transported to levels now below sea level on Starvation Canyon and Hanaupah Canyon fans (pp. 78-79), Death

Valley probably was above sea level and had exterior drainage to the Mojave River at Soda Lake, as already noted. These relative positions would not have been established by extensive earth movements; about 1,500 feet of downwarping of Death Valley in 500,000 years would have sufficed, and this amount is less than the rate of displacement during the past 2,000 years. The Pliocene Furnace Creek Formation was deposited in a playa, but not necessarily below sea level. Great Salt Lake Desert, for example, a larger salt-crusted playa than Death Valley, is 4,200 feet above sea level. It is likely that Death Valley has only recently dropped below sea level.

Of the 8,000 feet of fill in the valley, perhaps 3,000 feet is Quaternary and the lower 5,000 feet late Tertiary, like the Artist Drive and Furnace Creek formations. The valley has been forming all the time while these sediments were being deposited; the valley floor has been sinking and the mountains have been rising. The total vertical displacement therefore is about 4 miles.

The valley consists of three structural basins (see p. xii). The deepest one, with 8,000 feet of fill, is at the north under Mesquite Flat. A second basin, to the southeast at Cottonball Basin, is separated from the basin at Mesquite Flat by a structural uplift marked at the surface by Salt Creek Hills, where the Pliocene Furnace Creek Formation and overlying Quaternary gravels have been domed (fig. 14). The bedrock bottom of Cottonball Basin extends southeastward to form a southeast-trending structural valley along Furnace Creek Wash, separating the Black Mountains and the Funeral Mountains; a branch extends southward under Middle Basin.

The third deep basin under the floor of Death Valley is at Badwater Basin. The deepest part of the fill here is on the west side of the salt pan along the Amargosa River, almost opposite Bennetts Well. This basin is separated from Middle and Cottonball basins by a structural arch across the valley opposite Artist Drive. The continuity of this arch, discovered by drilling, is indicated by gravity surveys and by the recent 18-inch arching of the floodplain (see p. 122) which has isolated the floodplain in Middle Basin from the one in Badwater Basin.

These structural basins began to form in Oligocene time when the Titus Canyon Formation was deposited. Each one probably began as a series of small individual structural basins, each trending northwest. As the small basins became deeper and the bordering mountains became higher, the basins joined. One of the first basins in Oligocene time extended northwest across the north end of the Funeral Mountains. Three other basins, probably of about the same age, include one trending northwest from the cove at Badwater, one trending northwest from the cove south of Copper Canyon, and one along the Amargosa River at Mormon Point. All four basins were bordered by northwest-trending anticlines of Precambrian crystalline rocks exposed by the removal of Paleozoic rocks forming the upper plate of a flat fault thrust westward across the Precambrian rocks. The upper plate of Paleozoic rocks is still preserved at the Grapevine Mountains, the south end of the Funeral Mountains, and at Tucki Mountain.

FAULTS, GRANITE INTRUSIONS, AND VOLCANISM

The Oligocene basin deposits overlapped the anticlinally folded Precambrian rocks, but as the anticlines continued to rise and the basins continued to sink, the Oligocene deposits were dragged upward and slid off the flanks of the anticlines (fig. 97). The bare surfaces now exposed in the smoothly rounded Precambrian rocks above Badwater and Copper Canyon are referred to as turtlebacks (fig. 98). Where the turtlebacks are overlapped by Miocene formations, these formations too are dragged upward and have slid off the increasingly steep flanks of the anticlines. The basins that developed into the fault valleys may have started as downfolds.

FIG. 97. Oligocene formations (O) overlapped Precambrian rocks (pЄ) at foot of Funeral Mountains and were later downfaulted into Death Valley as mountains continued to be uplifted.

FIG. 98. Turtleback at Copper Canyon. The mountain of Precambrian gneiss (pЄ) has been raised by anticlinal uplift, and Tertiary formations (T) have been faulted by sliding off north (left) flank of anticline. Both Precambrian and Tertiary formations are displaced by block fault along front of Black Mountains.

This kind of structural relationship could be interpreted to mean that the displacements are all post-Miocene, but even the Funeral Fanglomerate is locally involved. Almost certainly the downfaulting of the valley and the uplift of the mountains have been progressing in small increments ever since the Oligocene, and the large north-south valley has evolved from a series of smaller northwest-trending ones. A recent paper (Wright et al., 1974) attributes the turtlebacks and other northwesterly trending structures to crustal extension.

Pre-Oligocene structural relationships are obscure, yet some major episodes are recorded by the relationships between the various rock formations. For example, on the east side of the Panamint Range, 3 miles north of Trail Canyon, a mass of volcanic rocks overlaps the Paleozoic formations (fig. 99). The volcanic rocks probably are Miocene or older. They dip 20 degrees to the east; the Paleozoic formations beneath them dip 45 degrees to the east. Since those volcanic rocks formed, the Panamint Range has been tilted 20 degrees to the east, and something like half the uplift of the Panamint Range relative to Death Valley has also occurred since then.

The Paleozoic and late Precambrian formations form the upper plate of a thrust fault known as the Amargosa thrust (fig. 100). The upper plate is widely broken by faults (fig. 101). The main fault differs from ordinary thrust faults in having younger rocks (mostly Paleozoic) thrust over older rocks (Precambrian) (fig. 102). At Tucki Mountain the upper plate is divided by a series of branching faults which curve upward toward the east and join the main fault at a flat angle. These branching faults, like the main one, also have younger rocks thrust over older ones; here, younger Paleozoic formations have moved westward over older Paleozoic ones.

The turtleback surface on the Precambrian rocks at the north end of the Funeral Mountains (fig. 97) is the exposed domed surface of the Amargosa thrust fault; the Paleozoic formations at the south end of the Grapevine Mountains and of the Funeral Moun-

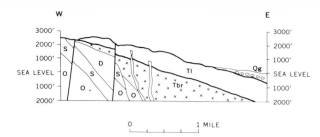

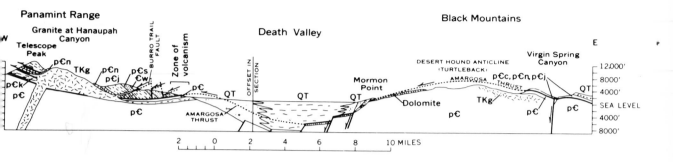

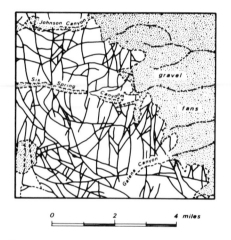

FIG. 99. Section along ridge 3 miles north of Trail Canyon. Tertiary lavas dipping 20 degrees east overlap more steeply dipping Paleozoic formations. About half the eastward tilt of Panamint Range has occurred since lavas were erupted. O, Ordovician; S, Silurian; D, Devonian; Tbr, Tertiary volcanic breccias, in part intrusive; Tl, felsitic lavas of Tertiary age; Qg, Quaternary gravel. (From U.S. Geol. Survey Prof. Paper 494-A.)

FIG. 100. Section of Black Mountains and Panamint Range showing supposed extent of Amargosa thrust and relations of granitic intrusions and volcanism to it (section across Black Mountains from Noble, 1941). Qt, Quaternary and Tertiary valley fill; TKg, granitic intrusions; €w, Wood Canyon Formation; p€s, Stirling Quartzite; p€j, Johnnie Formation; p€n, Noonday Dolomite; p€c, Crystal Spring Formation; p€k, Kingston Peak Formation; p€, Precambrian metamorphic rocks. (From U.S. Geol. Survey Prof. Paper 494-A.)

FIG. 101. Fracture system in upper plate (area without stippling) of Amargosa thrust in southern part of Panamint Range. Here it is composed of brittle late Precambrian formations, which extend under gravel fans and presumably are equally faulted there.

tains are remnants of the upper plate draped over this Precambrian turtleback.

Under the thrust fault in many places is chaos—and geologists call it just that. The Chaos consists of a mixture of brecciated and crushed Precambrian, Paleozoic, and Tertiary rocks. Some blocks in the mixture, shaped like elongate teardrops, have been referred to as whales. The blunt ends face toward the west. Some of the Tertiary rocks incorporated in the Chaos are crushed masses from formations older than the Chaos; some are igneous rocks intrusive into it (fig. 103).

Understanding the Amargosa thrust fault is next to impossible, but we can examine its parts. One of its branches has been named the Burro Trail fault because its outcrop follows the contour and has provided a favorite route for burros in that part of the mountains.

The Burro Trail fault can be followed from south of Hanaupah Canyon to north of Trail Canyon. Northward the outcrop of the fault cuts obliquely from late Precambrian rocks to late Cambrian formations. In Hanaupah Canyon (fig. 104) the fault dips 15 degrees west; beds of the early Cambrian Wood Canyon Formation

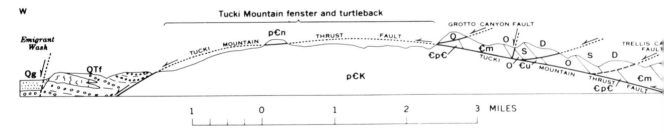

FIG. 102. Section of Tucki Mountain showing Tucki Mountain thrust fault and its branches on east side of mountain, turtleback on west side, and the Funeral Formation in Emigrant Wash which overlapped the turtleback and later was faulted down against it. pЄk, Kingston Peak(?) Formation; pЄn, Noonday(?) Dolomite; ЄpЄ, Stirling Quartzite and Lower Cambrian; Єm, Middle Cambrian; Єu, Upper Cambrian; O, Ordovician; S, Silurian; D, Devonian; QTf, Funeral Formation; Qg, Upper Pleistocene fan gravel. Vertical scale not exaggerated. (From U.S. Geol. Survey Prof. Paper 494-A.)

FIG. 103. Chaos at mouth of Death Valley Canyon. Monzonite porphyry in Chaos is intruded by dikes.

FIG. 104. View south along Burro Trail fault in Hanaupah Canyon. East-dipping Wood Canyon Formation (Cambrian) has moved westward across east-dipping Johnnie Formation (Precambrian).

in the upper plate dip eastward into the fault. In the lower plate are east-dipping beds of the late Precambrian Johnnie Formation. This segment of the fault is readily accessible by four-wheel-drive vehicles.

Northward from Hanaupah Canyon the fault cuts into upper Cambrian carbonate formations, and at the mouth of Death Valley Canyon it displaces the east-dipping limestone and dolomite beds. In the lower part of Trail Canyon, which is readily accessible except by low-slung modern automobiles, the fault carries Ordovician rocks westward onto Cambrian formations (fig. 105). This stretch of the fault is flat or dips 15 degrees west. As already noted, this part of the mountain was tilted 20 degrees east after the Miocene(?) volcanics overlapped the Panamint Range north of Trail Canyon. The Amargosa fault and its principal branches are older than the volcanics. Prior to the deposition of those volcanics, therefore, the Burro Trail fault and its parent, the Amargosa thrust fault, must have dipped west 20 degrees or more.

Before speculating on how the Amargosa thrust fault was formed, we must look at one more of its igneous features. In the headward parts of Hanaupah and Starvation canyons and at Harrisburg Flat, masses of granite rock intrude into the older rocks around them. These intrusive masses, which are eastern satellites of the complex of granitic intrusions forming the Sierra Nevada batholith, spread along flat faults of the Amargosa thrust system (fig. 106). They are tabular intrusions whose shapes were controlled principally by the flat faults of the area.

Since my fieldwork, and since this summary report was written, the intrusion in Hanaupah Canyon has been mapped and described in detail (McDowell, 1974). The more recent mapping shows the intrusion is much more irregular than indicated in figure 106 but confirms that the intrusion was physically injected and bulged eastward from a cross-cutting western source.

If my reconstruction of structural events is approximately correct, we may visualize a set of flat faults (Amargosa system) dipping more than 20 degrees west, which is about the slope of the present base of the crust in the area. The Moho, the boundary between the light crustal rocks and the denser rocks of the

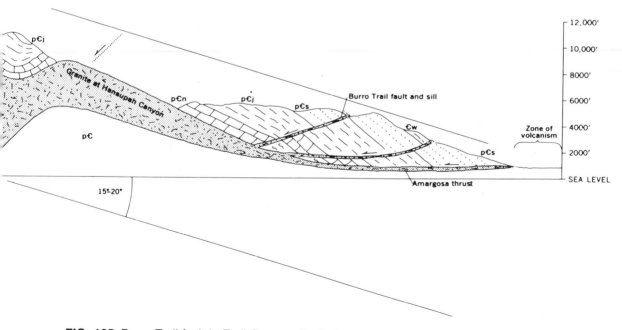

FIG. 105. Burro Trail fault in Trail Canyon. Faulted across upturned beds of Bonanza King Formation is a nearly horizontal slab of Nopah Formation.

FIG. 106. Idealized cross section of granite at Hanaupah Canyon. Intrusion is tabular with roof and floor of older rocks, but contacts are much less regular than indicated in diagram. Length of section, 8.5 miles; vertical scale not exaggerated. pϾ, Precambrian metamorphic rocks; pϾn, Noonday(?) Dolomite; pϾj, Johnnie Formation; pϾs, Stirling Quartzite; Ͼw, Wood Canyon Formation. Since faulting and intrusion of the granite, Panamint Range has been tilted 15 to 20 degrees east. (From U.S. Geol. Survey Prof. Paper 494-A.)

mantle, dips from Death Valley about 30 degrees westward under the Sierra Nevada.

Faults of the Amargosa system, dipping west and having granitic rocks intruded along them, are probably related to the growth and development of the granitic batholith in the Sierra Nevada. The main events at the batholith are older, but what can be seen in Death Valley may be the last manifestations of that batholithic activity. The vast amount of crustal rocks that had to be melted to form the batholith may have been supplied in part by rock masses sliding down the flanks of the furnace. The upper plate of the Amargosa thrust, the youngest of the suggested slides, never moved that far; the activity ended. The rocks along the faults, though, did become heated and in places melted and recrystallized (fig. 107). Masses of melted rocks were squeezed upward to form the intrusions and in places broke through the surface to contribute to the volcanics. Perhaps the melted rocks helped lubricate the faults.

The westward destination of the upper plate is obscure, and its source to the east is equally so. On Tucki Mountain, for example, 25,000 feet of Paleozoic formations, ranging from Cambrian to Permian, are in order except for shingling by the branching faults (fig. 102). Such a section could not be derived from a strongly folded and faulted source; the source must have been an area of gentle dips. The section could have been derived from the western edge of the Colorado Plateau at a time when the gently dipping plateau structure extended much farther west than it does now. Unless there was an escarpment of Paleozoic rocks 25,000 feet high, which seems unlikely, the fault that broke off the western edge of the plateau dipped westward into the ground, as postulated for the Amargosa fault system.

The source area with gentle dips must have been sufficiently far west of the present edge of the Colorado Plateau for the Paleozoic formations to thicken from 5,000 feet on the plateau to 25,000 feet in the source area of the plates of thrust-faulted rocks now on Tucki Mountain. These Paleozoic formations continue to thicken from Death Valley westward toward the Sierra. Below the Paleozoic was another 25,000 feet of Precambrian sedimentary formations, which rested on the crystalline basement. The surface of the crystalline basement must have had considerable slope to the west, greater than 10 degrees, and when the western support for the thickened section was weakened by igneous activity at the batholith, the mass could begin sliding west (fig. 108). The Precambrian sedimentary rocks could slide westward onto older Precambrian crystalline rocks, and the Paleozoic could slide westward onto the late Precambrian rocks.

Many, perhaps-most, geologists have doubts about this tentative explanation of the Amargosa thrust fault. I must acknowledge that I do too, but so far as I know nobody else has even made an attempt to explain the fault.

FIG. 107. At some places along Amargosa thrust fault, rocks in Chaos melted and recrystallized. Masses of older rocks (black) were engulfed and partly absorbed by melted rocks (light).

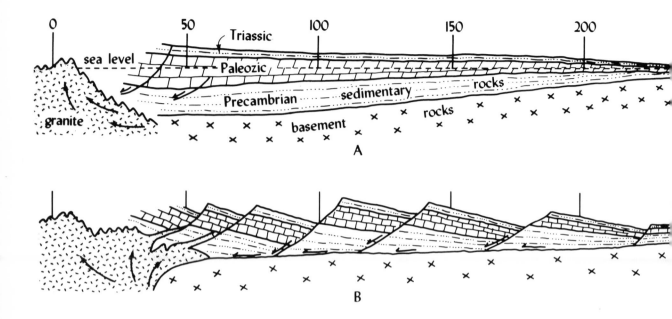

FIG. 108. Attempt to understand the Amargosa thrust fault. *A.* Late Cretaceous: Precambrian, Paleozoic, and Triassic formations thicken westward
and slope toward granite (batholith) of Sierra Nevada. As granite moves
upward, support for westward-sloping formations is removed and they begin
to slide westward. *B.* Early Tertiary: As sliding extends eastward, upper
plate of geosynclinal rocks becomes broken by faults that branch upward
from main fault on surface of basement rocks. Sheetlike masses of granite
spread eastward along faults, further lubricating them and facilitating sliding.
C. Middle Tertiary: As lifting continues, basement and overlying faulted
geosynclinal rocks become faulted. Original faults and faulted blocks are
tilted eastward; faults on which sliding occurred approach horizontality;
faulted sedimentary formations dip steeply east. New granite, generated in
basement rocks, intrudes upward along faults and erupts at surface. *D.* Late
Tertiary–Quaternary: Already complex Middle Tertiary structures are broken
by high-angle block faults, increasing eastward tilt; original faults on which
sliding occurred dip east; Precambrian and Paleozoic formations dip steeply
east.

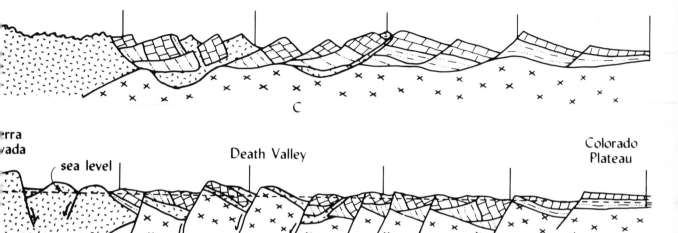

C

Sierra
Nevada

sea level

Death Valley

Colorado
Plateau

D

7   Mines and Mining

PRESENT PROBLEM IN DEATH VALLEY

When Death Valley became a national monument in 1933 it was
left open for prospecting and mining, thus posing many problems
for those charged with the responsibility of protecting the area for
use by all. No satisfactory solution to the problem has yet been
found. The solution must lie in the middle ground between the
extreme view that mining and prospecting may continue unre-
stricted, regardless of how these activities deface the landscape,
and the opposite view that land set aside as a park or a monument may have no productive uses.

   One solution might be to evaluate potential mineral resources
in parks and monuments as percentages of the national production of each particular commodity. A mineral deposit may be of
substantial value to its owner yet not be important in terms of
national production. Such deposits should not be developed on
withdrawn lands; the owner should be reimbursed but the land
should be protected for its scenic or other values. Deposits large
enough to be considered a factor in national production should
be developed and the scenic or other values compromised for the
material product. Because of built-in biases, opinions will differ

about the percentages to be accepted as cutoff limits, but at least such a formula would enable even extremists to come to a bargaining table.

Throughout the West, abandoned mines possess a certain romantic fascination for tourists, even though they may have been opened primarily to promote the sale of stock. Death Valley is no exception, and even a park ranger's eyes brighten at the mention of a name like the Garibaldi mine. Yet an active, productive mine is regarded as lower than a burro. It is true that some mines presently being worked in the monument are unkempt, but with a little planning and a lot of cleanup the talc mines could be made into a tourist attraction and an educational asset. For this purpose to be achieved, attitudes will have to change, among both the mining people and the Park Service. At this writing, both sides have assumed an unreasonable posture.

MINING HISTORY

The history of prospecting and mining in and near Death Valley may be divided into four main periods. An early period of prospecting for gold and silver lasted until about 1877, when a boom in the Panamint district ended. A second period began in the 1880s when borax was discovered (figs. 109, 110). The third period began in the first decade of the twentieth century, when interest in silver and gold revived. Skidoo and Chloride Cliff were productive, and excitement arose over the possibility of finding copper at Greenwater. The fourth period began after World War II, when production of talc became a major activity.

Early Mining   Among the early parties of prospectors were those seeking the lost and probably legendary Gunsight and Breyfogle mines. Like all parts of the West, Death Valley has its lost mines, and they seem to provoke more activity, on the part of both pros-

FIG. 109. "Bloom" of cotton balls (ulexite, or sodium calcium borate) on floodplain in Cottonball Basin, site of earliest mining for borax.

FIG. 110. Although borax was first produced in Death Valley from the salt pan, prospecting in nearby hills led to discovery of vein deposits of colemanite and other borate compounds. Since these deposits are richer than those on valley floor, the center of production gradually shifted to the hills. View shows adits on Monte Blanco, at head of 20 Mule Team Canyon.

pectors and authors, than mines that merely produce minerals.

One legend of early prospecting in Death Valley, about the Garibaldi mine at the head of Tucki Wash, seems so incredible that few have believed it and no one has written about it. I too was incredulous about the account when I first heard it, from Ranger Matt Ryan of the Park Service, but I was later able to confirm and document the story.

The mine was visited by the Wheeler Survey in the 1870s and shows on the map made by the group. Shortly after the visit by the Wheeler Survey, the owners quarreled and both were killed. When I first saw the cabin at the mine I found it just as Ryan had described it. Two skeletons faced each other across the table, both slumped forward. The right arm of each was extended; each had an old-time pistol in his right hand. Each had been shot by the other, in exactly the same place. There is a bullet hole through the seventh rib in front and the ball is embedded in the seventh rib directly behind it at the back. Both skulls face left. In the middle of the table between the two is the loot about which they had quarreled: a pile of gold coins, a diamond necklace, and other jewelry. It was obviously more than had ever been produced from the Garibaldi mine. A person of low-character would have expected to pocket some of the coin and jewels, but in a national monument visitors do not collect plants or stones. The jewels, and tall tales like this one, are left for the enjoyment of all.

When the Panamint district boom ended in 1877, interest in metal shifted from the Death Valley area westward to Darwin, and for a few years Death Valley was rarely visited.

Borax   Meanwhile, in other parts of the West, interest in industrial chemicals was growing. The discovery of borax ushered in the second period of Death Valley mining history. Borax had first been found, in the American West, at Borax Lake, in Lake County, California, where production started in 1864. During the 1870s borax was produced at Big Soda Lake, Columbus Marsh,

Teels Marsh, and other borate playas in Nevada and at Searles
Lake and in the Saline Valley in California. When production
started in the United States, the country was importing $200,000
worth of Borax annually; soon, with domestic production flour-
ishing, imports declined sharply and the price of borax fell from
50 cents to 30 cents a pound. By 1907 it had dropped to 5 cents a
pound. Death Valley, though one of the later areas to be exploited, figured prominently in the history of borax, even after it
ceased to produce the commodity, because of the well-advertised
20-mule teams which, in the early days, made the haul from the
valley to Mojave.

The history of borax production in Death Valley began in 1881,
when Aaron Winters prospected Cottonball Basin and Isadore
Daunet prospected Badwater Basin east of Eagle Borax Spring.
The Eagle Borax operation was soon abandoned, but Winters
sold his claims to William T. Coleman, who had the financial
resources necessary to develop the properties.

Borax was first produced in Death Valley from the floor of
Cottonball Basin where fibrous nodules, the cotton balls (fig. 109)
of ulexite and probertite (both sodium and calcium borates), were
gathered in baskets by Chinese coolies. The cotton-ball ore was
treated in the Harmony Borax works by a method not exactly
known, but apparently it was mixed with an equal quantity of
sodium carbonate and the mixture was boiled for about an hour.
Insoluble matter would settle to the bottom of the vat and calcium carbonate would precipitate from the clear liquid as it
cooled; the remaining liquid, on evaporation, would yield borax
(sodium borate). Sodium carbonate could have been gathered at
the marshes 2 to 4 miles north of the Harmony plant, but there
may have been nearer deposits that are now worked out.

Within a year much richer deposits of colemanite (calcium
borate) were found in Tertiary formations (fig. 110) in the northern part of the Black Mountains, notably at Monte Blanco and
farther east at the Lila C. mine and near Shoshone. In the late

1880s Death Valley production was interrupted when rich deposits were found at Borate, in the Mojave Desert, only 6 miles from the Santa Fe Railroad.

Borax mines and prospects on the salt pan, where production started, include conspicuous mounds, furrows, and other kinds of workings, but they were not used in the early activity. At first, the cotton balls were simply collected from the surface of floodplain areas of the salt pan (fig. 109), placed in bushel baskets, and hauled to the Harmony Borax works for conversion to borax. The operation left no mine workings; only field camps and the borax works remain.

The conspicuous mine workings on the salt pan date from the period 1900-1910, when the land was patented and the borax company had to show developmental work on the project. A veteran company engineer who had participated in the operation told me that the instructions were simply to "Pile it high." These working are of three kinds. Some are parallel furrows closely resembling a coarsely plowed garden; good examples are in the smooth silty rock salt close to the paved road at the parking area near Mushroom Rock. The second kind (fig. 111) consists of mounds arranged in rows; they are abundant on the floodplain 2 miles northwest of Furnace Creek Ranch. One must walk to them. The third kind of workings consists of large solitary mounds at marshes. A group of them can be seen at the marshy northeast corner of Cottonball Basin, another at the marsh on the west side of that basin, and a third at Eagle Borax on the west side of Badwater Basin.

Later History   The third period in Death Valley mining history began in the first decade of the twentieth century when gold was discovered at Skidoo, Chloride Cliff, and Rhyolite (a major camp 25 miles north of Death Valley), and the possibility of finding copper at Greenwater, on the east side of the Black Mountains, raised fervent hopes. These developments came during a lull in

FIG. 111. Parts of floodplain which are frequently flooded are bare mud flats. The mounds are part of the work that was done for patenting the land in the period 1900-1910. They originally were small heaps of mud, but sodium chloride was deposited by capillary rise of moisture. View is southeast in Cottonball Basin.

the mining of borax in Death Valley, but the camps began to
close in 1907 because of the countrywide financial panic.

The amount of gold and silver produced at camps in Death
Valley is a matter of guesswork because production information
was withheld. It seems doubtful, however, that gross production
from all the camps aggregated more than a few million dollars.
Two construction jobs are of interest in connection with mining
camps. A telephone line was built between Skidoo and Rhyolite,
a distance of almost 50 miles; a buried pipeline almost 20 miles
long brought water to Skidoo from springs in the vicinity of Telescope Peak.

After the gold camps began to close in 1907, metal mining in
Death Valley dwindled; by 1910, for all practical purposes, it had
ended. In a flurry of prospecting during the 1930s many people
sought mineral deposits as a means of eking out a subsistence.
Metal production in the Death Valley area has never been
significant, and consequently mining for metals has been the
result rather than a contributing cause of economic conditions in
the rest of the nation. Not so with the less glittering borax. In the
midst of the panic, as metal mines were closing, the Lila C. mine
resumed production. Borax mining in the Death Valley region
strongly influenced the national economy, as did the later talc
mining.

The Lila C. mine was a major producer of borax for seven
years, until 1914 when the properties at Ryan were developed and
a narrow-gauge railroad was extended to reach that town. During
the period when the Lila C. and Ryan mines were productive, the
Death Valley area produced most of the borax consumed in the
United States, perhaps as much as 80 percent. In 1928 operations
at Ryan were discontinued, and since then there has been very
little borax produced in Death Valley.

Production of soapstone and talc (fig. 68) in the region began
during World War I. The principal deposits are located in the
Crystal Spring Formation of Precambrian age which outcrops in a

discontinuous belt from the south end of the Panamint Range southeastward to the Kingston Mountains. By 1960 this belt was producing about 25 percent of the nation's talc.

During World War II, the beginning of the fourth period of Death Valley mining history, salt was mined at the deposit of massive rock salt north of Badwater for use at a magnesium plant near Las Vegas, Nevada. According to Park Service records, about 15,000 tons of salt had been produced by August 1942, when rains and flash floods destroyed the road leading out of Death Valley. Production was not resumed. This operation is of interest for providing a measure of the rate at which salt pinnacles develop on the deposit; after twenty years the smoothed salt surface, covering about 20 acres, has roughened only slightly. There are no high pinnacles on that part of the salt surface.

Some carbonate may have been produced in Death Valley in the 1880s to operate the vats for crystallizing borax at the Harmony works.

In the sulfate zone around the edge of the salt pan in Death Valley are deposits of gypsum, but they have not been exploited. When heated, gypsum ($CaSO_4.2H_2O$) loses much of its water and changes to a hemihydrate ($CaSO_4.\frac{1}{2}H_2O$). This calcined gypsum, known as plaster of Paris, combines with water to set and become hard. These properties and uses of gypsum were known to Egyptians as long ago as early dynastic times.

Epsom salts, the hydrated magnesium sulfate used chiefly in medicine, is widely distributed in Death Valley but is found mixed with other salts. The salt was named for Epsom, England, whose mineral springs made it a fashionable spa as early as 1618.

Nitrates in Death Valley were the object of intensive prospecting during World War I when the United States needed additional supplies of this commodity. No important deposits were discovered, but the demand was met by the development of synthetic sources. Nitrates have long been sought because of their uses in fertilizers and explosives. Until World War I, Chilean deposits, discovered about 1830, were the world's principal source of nitrates.

Nitrogen salts were known to the ancients too, but the one known to them seems to have been sodium nitrate and not saltpeter (Lucas, 1948:304). Moreover, according to Lucas, the word "niter," which now means saltpeter, seems to be a mistranslation where it appears in ancient literature: for example, "as vinegar upon nitre, so is he that singeth songs to an heavy heart" (Prov. 25:20). Since, as noted by Lucas (ibid.), saltpeter does not react with vinegar, the passage in Proverbs probably refers to sodium carbonate (natron) which dissolves in vinegar with effervescence.

Potash salts were also in short supply during World War I and, like nitrates, were the object of intensive prospecting in Death Valley. The U.S. Geological Survey bored several holes seeking potash brines in Badwater Basin, and the Pacific Coast Borax Company drilled three deep holes. The Death Valley search was not successful but domestic sources were found at nearby Searles Lake and in the Permian basin of New Mexico and Texas.

Potash literally means "pot ash," the material left in a pot by evaporation of a solution containing wood ashes. Ashes of sea plants contain principally sodium carbonate; ashes of land plants contain potassium carbonate. The ancients knew that salts could be obtained in this way. Aristotle (ca. 340 B.C.) describes the method (*Meteorology*, II, iii, 161): "In Umbria . . . there is a place where reeds and rushes grow; these they burn and throw their ashes into water and boil it till there is only a little left, and this when allowed to cool produces quite a quantity of salt."

For the next 2,000 years there was little if any improvement in the method described by Aristotle. Production of alkalies in Europe led to forest depletion; an acre of woods would yield 2 tons of potash, and three men could annually cut and burn 12 acres. Early in the seventeenth century attempts were made to start production in colonial America, and in 1751 Parliament remitted duties on potash in order to encourage the infant industry. Premiums were offered for raising plants with the highest yield, such as borilla, glasswort, and other chenopods. The first American patent, signed by George Washington in 1790, authorized a process of producing potash and a refined product, pearl ash (*New York Times*, April 10, 1960).

The value of American production of potash and pearl ash reached 1.5 million dollars in 1809 and 1810. Production of potash from wood ended when the salt deposits at Stassfurt, Germany, were developed and the Leblanc process and, later, the Solvay process yielded a higher quality of soda ash than could be obtained from plant ash.

At the time of the earliest dynasties, natron (sodium carbonate) was produced in the Wadi Natrun, a valley in Egypt, where it occurs naturally as an efflorescence around dry lakes, probably like the deposits along the east edge of Cottonball Basin in Death Valley. Natron was used in Egypt as a dehydrant in mummification and possibly, when mixed with malachite and fired, as a component in Egyptian blue glaze (Lucas, 1948:198, 317).

The annual production of salt in the United States is now approximately 7 million tons, but only a small amount of the product, considerably less than 5 percent, is destined for household and table use. Most of it, probably three-quarters, is used in the chemical industry, particularly for the manufacture of sodium sulfate (salt cake), hydrochloric acid, chlorines, chlorates, and caustic soda (see table 3).

After World War II, Death Valley, like the rest of the nation, swarmed with uranium prospectors. During the 1950s there was a flurry of prospecting for tungsten east of Skidoo. Since World War II there has also been prospecting for perlite and for some of the colorful volcanic and other rocks to be used as decorative building stone. This kind of activity in Death Valley National Monument, however, is unwarranted, for there is no shortage of such materials in other parts of the United States.

8 Archaeology of Indian Occupation

The archaeological record indicates that there have been Indians in Death Valley during most or all of Holocene time (the past 10,000 years) and perhaps as long ago as late Pleistocene. Yet even in the halcyon days when springs were more numerous and larger than they are now and when the floor of Death Valley was a lake instead of a salt-crusted mud flat, the Indian population appears never to have been large.

CHRONOLOGY

The prehistory of Death Valley may be divided into four stages. The earliest stage, characterized by a type of projectile point called Lake Mojave (fig. 112), may be as old as late Pleistocene. There was more water then; there may have been a lake in Death Valley as there was at Soda Lake (Pleistocene Lake Mojave) where the traits characterizing this occupation were first recognized.

During the early Holocene, Death Valley became drier and hotter than it is today. It is doubtful that Indians visited the valley more than rarely. With the approach of the middle Holocene (ca. 3000 B.C. to A.D. 1) the climate again became milder and wetter, a shallow lake about 30 feet deep formed on the valley floor (see chap. 2), and the valley again was occupied by Indians who represented the second recorded prehistoric stage and may have overlapped the end of the first occupation. These Indians continued to use the spear, and the atlatl rather than the bow and arrow, and to rely mainly on hunting. Some springs they used are now dry, showing that the climate then was wetter. This recent occupation in Death Valley may have flourished about the time of the Egyptian dynasties that built the great pyramids.

FIG. 112. Atlatls or spearpoints from earliest Death Valley sites. Points are of Lake Mojave type, except for center one in top row which is of Silver Lake type. Average weight is 8 grams; arrowpoints, used thousands of years later, usually are less than a quarter the weight of these projectile points. Rounded ends of some points suggest they may have been used as knives. Specimen in lower left is 5.25 centimeters long.

Sometime after A.D. 1, when the middle Holocene lake had dried up, the third group of Indians came into the valley and brought with them the bow and arrow. Five hundred or a thousand years later the fourth occupation introduced pottery and a new type of arrowpoint. The latest valley Indians, and perhaps those of the third stage, were ancestral to some of the Shoshone speaking tribes that were in the region when the first whites arrived.

DEATH VALLEY I INDIANS

The earliest signs of Indians in Death Valley are campsites on gravel benches near springs. These benches have smooth surfaces known as desert pavement (fig. 113; see also chap. 4), in which stone tools left behind by early campers are embedded. Since both the tools and the pebbles and cobbles forming the pavement are darkly stained with desert varnish, the tools are difficult to find. One site has half-buried rocks in two semicircles about 10 feet in diameter, around which tools are concentrated. These semicircular rings probably are remains of shelters. Large boulders on the sites as well as rocks piled along trails leading to springs or overlooking them may have served as hunting blinds. Perhaps hunting companions drove the game—mountain sheep—toward the springs, as was done in historic time.

Part of the evidence that there was more water at the time of the earliest Death Valley Indians is along Furnace Creek Wash where there was a spring, now dry. Rounded groups of rocks embedded in the desert pavement at this site probably were fireplaces (fig. 113), although no charcoal remains near or under the rocks. The rocks are selected types, mostly one kind and approximately one size, and many are fractured as if fired. Tools tend to be concentrated in the area of these supposed fireplaces.

This site was also used by Death Valley II Indians, but there are few signs of later camping here although the main foot trail into Death Valley passes near the location and an old wagon road crosses it. No Death Valley III or IV artifacts (with one possible exception) were found. There are no tin cans or bottles (see chap. 9). Evidently the spring was flowing during the wetter climatic epochs in Death Valley, but it seems to have dried up about 2,000 years ago.

Almost two-thirds of the artifacts found on Death Valley I sites are scrapers (fig. 114), used for removing hair and flesh from hides. Scrapers made like these are not found on later sites in Death Valley. Other tools—knives (fig. 115), perforators, and a few big choppers—are also found on old sites. A distinctive kind of rock was commonly used, especially for scrapers—a brown chert from Paleozoic limestone or dolomite formations, or silicified volcanic rock weathered to a deep golden color.

Other early-type projectile points have been found in Death Valley, some on old travertine that once was swamp. Although possibly related to the Death Valley I occupation, they were found as isolated items and not as parts of a tool complex.

FIG. 113. Probable fireplace at early Death Valley site along Furnace Creek Wash, near a now-dry spring. Fireplace stones are mostly dolomite, although desert pavement is formed of many different kinds of stone. Tools are concentrated around the fireplaces. Dolomite stones are angular; many seem to be broken, as if they had been in fire. Similar concentrations of stones are not found on gravel benches away from known early sites.

FIG. 114. Early Death Valley round and semicircular scrapers. Specimen in lower left is 5 centimeters wide.

FIG. 115. Death Valley I knives, many of them made of brown chert. Specimen in lower left is 3.5 centimeters wide.

Tools of these early Indians suggest that they subsisted largely by hunting. No grinding stones were found at their sites, but surely they gathered nuts, berries, and greens. The game provided meat, and it could have been well salted. The climate, even in a wetter time, would favor dried meats, or jerky. Skins must have been used for protection against the wind; the Indians may also have used hides for footwear and carrying bags. Presumably, like the historic Indians, the early inhabitants escaped the summer heat by moving into the mountains.

DEATH VALLEY II INDIANS

Whereas the known sites of the earliest Indians are mostly in desert pavement on gravel terraces, Death Valley II sites are largely on soft silty ground. One such site, near the south end of the salt pan, is near a spring with water that, though now too salty to drink, was used both by Death Valley II and by Death Valley III prepottery Indians. The water has not been used, however, during the past thousand years, probably because its salinity has increased with the decreased flow of water.

The Death Valley II site at a dry spring along Furnace Creek Wash which was also used by earlier Indians centers at a large, low mound of boulders (fig. 116), probably a burial place although no bones were found. Excavation of this mound revealed an oval pit 15 inches deep, measuring 4 by 5 feet. Projectile points and other tools found in the mound are large and well made (figs. 117, 118). The largest projectile point is almost 4.5 inches long and weighs more than 26 grams. The Indians continued to hunt with spears and atlatls, but the points are quite different from those of the earlier Indians.

There are other differences. On our Death Valley II sites knives are twice as abundant as scrapers, whereas the earlier sites have four times as many scrapers as knives. The materials used for

FIG. 116. Mound at Death Valley II site along Furnace Creek Wash was built of about 200 boulders, ranging up to 18 inches in diameter, all darkly stained with desert varnish. Some were broken by cross fractures after being placed on mound. The boulders filled a pit 5 feet in diameter and 15 inches deep. Excavation disclosed artifacts shown in figures 117 and 118. These artifacts, and others illustrated in this book, are in the Death Valley Museum.

FIG. 117. Death Valley II projectile points recovered from site shown in figure 116. Specimens second from left in top row and second from right in bottom row show white coating of calcium carbonate. Such coatings form quickly in wet environments, but in dry gravel, as at this mound, they form slowly. Specimen in lower left is 10 centimeters long.

FIG. 118. Six knives recovered from Death Valley II site shown in figure 116. Specimen on left is 13.5 centimeters long.

making tools also differ. Death Valley II tools are usually made of chert from volcanic rocks. Black basalt also is common and was not used for Death Valley I tools.

The later Death Valley II Indians used corner-notched projectile points, but these, too, are large and most likely were used with atlatls. Many knives and drills were found with them. The traits characteristic of the late Death Valley II Indians are not well known from this survey because their tools were found only at mixed sites around the salt pan. Evidently, though, the middle Holocene lake was drying. The presence of a few grinding tools suggest that, as the springs dried, reliance for survival was shifting from hunting game to gathering seeds, nuts, and berries to be eaten raw or to be roasted and ground and made into cakes and gruel. Evidence collected elsewhere suggests that there may have been increased use of basketry.

THIRD STAGE—ARRIVAL OF THE BOW AND ARROW

Sometime after A.D. 1 the Indians in Death Valley changed from hunting with the atlatl to hunting with the bow and arrow, and the way of life that began with this change is referred to as Death Valley III. The most characteristic arrowpoints used by Death Valley III Indians are long and slender and average less than 2 grams in weight (fig. 119). Perhaps the change in technology was caused by a decrease in larger game, which forced the Indians to depend more on rabbits, wood rats, and other small animals. These Indians were largely nomadic and lived by hunting and gathering over a wide area, spending winters at springs around the salt pan and summers at higher altitudes in the mountains.

The larger number of metates (many portable), manos, and pestles found at Death Valley III sites show the increasing reliance on the gathering of wild seeds and plants. That the Indians began to live above the level of subsistence is attested by the presence of ornaments and unbaked clay figurines. They undoubtedly visited, and were visited by, or traded with their Basket Maker III, Pueblo II, and Virgin River neighbors to the east and northeast, and like their eastern neighbors they wore imported olivella and limpet shell beads probably carried there from the west by the traders of the desert, the Mojave. They also made pendants and unbaked clay figurines, and they buried their dead under rock mounds. Their arrowpoints are corner notched, and when made with care, like those found at burial sites, they are long, slender, and beautifully fashioned. Such points are characteristic of a wide area to the east of Death Valley.

Death Valley III Indians lived in the dunes at springs bordering the salt pan and along Salt Creek and the Amargosa River. Late Death Valley II tools are found in the alluvium or gravels under the dunes; Death Valley III tools are found at the lowest levels of the dunes or on playa beds near the dunes. The sites at the edge of the salt pan are on what had been the bed of the middle Holocene lake. Some of the dunes may have been started by Indians tramping the sandy ground.

FIG. 119. Death Valley III projectile points and shell beads. Specimen on left is 4 centimeters long.

One small Death Valley III house site was found in a fork of Hanaupah Canyon (fig. 120). Three flat or slightly concave metates and one subrectangular, uniface mano were lying against a boulder wall enclosing an area 7 by 10 feet adjoining a cliff and under a slight overhang. A 2-foot break in the boulder wall provided an entrance. Fire-blackened areas lay against the face of the cliff and toward the outer wall. In the center of the shelter was a small pit, 15 inches deep, narrowing from 27 inches in diameter at the top to 18 inches at the stone-lined base. The floor of the shelter consisted of an occupation layer 6 inches deep under an inch of sand and debris. In it were found two projectile points of Death Valley III type, two blue schist pendants, and fragments of pinon nuts and mesquite beans.

The archaeological survey revealed that Death Valley III Indians buried their dead in pits over which they erected mounds of rock similar to the one shown in figure 121. One infant's grave was under a mound of rock on the surface and not in a pit, but it probably dates from the Death Valley IV period. In contrast, another mound that yielded Death Valley III artifacts marked a pit almost 3 feet deep containing a double burial, a flexed adult and a small child. Funeral goods included eighteen Death Valley III arrowpoints, more finely flaked and longer than the typical hunting points. With them were bone tools, a bead of bird bone, and an **L**-shaped bone sickle(?). The pit was filled with cobbles built upward in a mound 2.5 feet high, estimated to total 5 tons in weight.

Death Valley III and later Indians also made rock alignments. Some, like the one illustrated (fig. 122), are elaborate; others are merely sinuous serpentlike lines of cobbles.

FIG. 120. Death Valley III house site in fork of Hanaupah Canyon. Two fire pits (hollows), together with projectile points and pendants, were uncovered. Seated figure is Alice P. Hunt, who made archaeological survey; standing figure is Thomas W. Robinson of the U.S. Geological Survey.

FIG. 121. Burial mounds are numerous on gravel fans near Death Valley III and IV sites. Death Valley III mounds are similar to Death Valley IV mound shown here.

FIG. 122. Near end of this rock alignment, which covers half an acre, resembles serpent's head and coils; at far end is an oval mound of rock.

DEATH VALLEY IV INDIANS—ARRIVAL OF POTTERY

About A.D. 1100 or a little later, some Death Valley IV Indians, a Shoshone-speaking people related to the Southern Paiute, entered the valley from the east or the south, camping mostly at water holes and springs both east and west of the salt pan. Many of these sites had been used by Death Valley III Indians, and they may even have been shared with Death Valley III Indians in a transitional period, like those at Lost City at Overton, Nevada, 100 miles to the east. The resulting cultural blend would explain the persistence of some traits, such as burial under rock mounds and rock alignments, into the Death Valley IV occupation. The visits of the Shoshone-speaking people to the water holes and springs were brief; the Indians sometimes carried cooking vessels with them. They continued to visit the valley floor sporadically into historic time. Probably much of their equipment was of wood and basketry which has disintegrated on surface sites.

Another group of Indians, the Panamint Shoshoni, frequently camped on Furnace Creek fan. They probably came from the north and the northeast, rather than through Ash Meadows or from the south. These Indians had many of the traits of the puebloid (Fremont) cultures of western Utah, the Virgin River drainage, and southern Nevada, cultures consistent with their nonagricultural, seminomadic hunting and gathering way of life. They were well adjusted to make the most of their desert environment, as evidenced by numerous traps for small game, blinds for large game, a wide variety of grinding stones, and mesquite storage pits, and they had enough surplus energy to decorate themselves with pendants and beads and to make objects of unbaked clay.

A third group of Indians who may have visited Death Valley, or at least traded pottery vessels there, were the Yuman Walapai to the east and south along the east side of the Colorado River. On the Amargosa Desert, the next valley east of Death Valley, their sherds are found associated with Pueblo types.

From Furnace Creek south, the historic Death Valley Indians were a mixture of Shoshone, Southern Paiute, and Kawaiisu, the latter an ancient offshoot of the Southern Paiute (Steward, 1938).

Death Valley IV sites are characterized by pottery and by small triangular projectile points (figs. 123, 124). Most of the pottery was utilitarian, unpainted, brown or gray in color, and undecorated except for rare finger marks below the rim, especially in Southern Paiute pieces. The centers of pottery making were east and south of Death Valley, and most of the pottery found in the valley came in from those directions. Only one type of ware (see fig. 123) is definitely known to have been made locally. These later Indians, like the earlier ones, made pendants of talc and schist (fig. 125), imported shell beads, and in historic time were traded glass beads (fig. 126).

The Indians established winter sites mostly in dunes around the salt pan, and in this mild climate they needed little more

FIG. 123. Death Valley IV pottery, of the kind known as Death Valley brown ware, which was made locally. It was tempered with schist from Precambrian Johnnie Formation. The pot is 12.5 centimeters in height.

FIG. 124. Death Valley IV projectile points with side and corner notches date from before white contact; others date from before and after white contact. Specimen in lower left is 3 centimeters long.

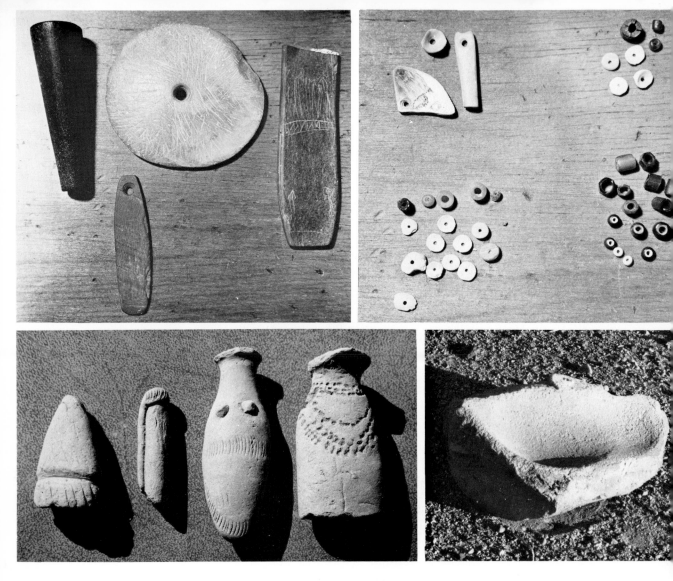

FIG. 125. Death Valley III and IV pendants and pipe. Pottery pipe (upper left) was imported from Lost City area in Nevada. Circular pendant is made of talc. Other two pendants are made of schist, probably from Johnnie Formation (see chap. 5). Pendant at right, decorated with figures of a brave and a maiden and with trails crossing on reverse side, has been referred to as "Death Valley's first love letter." Talc pendant is 6 centimeters in diameter.

FIG. 126. Pendants of bone or shell (upper left) and shell beads (large white) are found on Death Valley III and IV sites. Glass beads (mostly dark; two small white ones in lower right) were found on historic sites. Shell beads average about 0.5 centimeter in diameter.

FIG. 127. Figurines of unbaked clay from dunes in mesquite grove on Furnace Creek fan. Specimen on left is 3.5 centimeters long.

FIG. 128. Double-basin metate, about 50 centimeters long, from dunes a mile northwest of Bennetts Well.

than a windbreak or a wickiup, built of upright sticks and thatched. Some of these shelters were held down by rocks.

Objects of unbaked clay (fig. 127), resembling miniature carrying baskets, vessels, cylinders, figurines, and animal heads, are also found on Death Valley III and IV sites. These objects are quite like those made by the Indians at Lost City (A.D. 950-1100) on a tributary of the Virgin River in Nevada.

The change from a hunting to a food-gathering economy in Death Valley started in late Death Valley II. Grinding stones first appear on late Death Valley II sites, are more numerous on Death Valley III sites, and became still more numerous on Death Valley IV sites. The most common type of grinding stone, a rock slab with a flat or very slightly concave grinding surface, is called a metate; the stone held in the hand and rubbed against the metate is called a mano. Some metates characteristic of Panamint Shoshone Death Valley IV sites have two basins (fig. 128). Similar double-basin metates are found at tenth- and eleventh-century village sites (Fremont) in central and western Utah.

Food was also ground in mortars with pestles. The mortar might be of mesquite wood, like the one illustrated in figure 129, or it could be a grinding hole in a small boulder that could be carried, in a larger boulder, or even in bedrock.

Circular enclosures rimmed by gravel and rock are common on gravel fans near Death Valley III and IV campsites in mesquite dunes. In the course of the archaeological survey hundreds of such rock circles were found. Objects and structures whose purpose is obscure are commonly referred to as ceremonial, but many of the rock circles were taken out of that class when they were found to be pits used for storing mesquite beans and piñon nuts (fig. 130). Evidently the pits were dug on gravel fans for protection against rodents living in the dunes. Other circular cleared areas along trails crossing gravel fans were probably cleared sleeping areas for weary travelers. Some bolstered wickiups. A few were rolling places for animals.

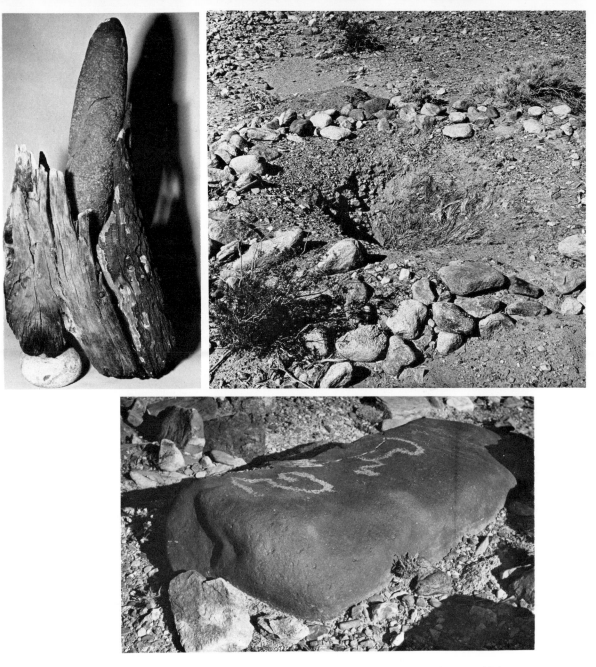

FIG. 129. Death Valley IV mortar and pestle, about 60 centimeters high. Mortar is hollowed mesquite wood.

FIG. 130. Excavated mesquite pit, roughly 1 meter deep and 1.6 meters in diameter, at rock circle 1.25 miles south of Bennetts Well. Pit had been lined with sacaton grass, part of which is still in place. Gravel wall of pit can be seen at left. Mesquite beans were stored in the grass, and the mat was then covered with stones and earth. (Excavated by Alice P. Hunt and Buddy Welles.)

FIG. 131. Petroglyph on boulder at foot of Hanaupah escarpment, west of Tule Spring.

A group of pits, probably for storing mesquite pods, built against a Holocene fault scarp on the east side of the valley (figs. 92, 93), helped to date that fault. Clearly the displacement occurred before the storage pits were dug.

For catching rodents and other small game, the Indians used deadfall traps. A figure-4 stick arrangement was set on a large slab of rock and supported an upper slab; mesquite pods were commonly used as bait. At one such deadfall were the bones of a desert wood rat and the mesquite pod bait trapped between the upper slab and the basal slab. More than fifty such traps of this kind were located by the survey; each is enclosed in a rock wall, perhaps to keep other animals out until the trap was revisited by its builder, or to keep the wounded animal inside.

The shoshone-speaking Death Valley IV Indians hunted bighorn sheep as well as rabbits, and something about the animal geography of the past thousand years or so is suggested by the location of these sites. Today, sheep are numerous on the east side of the valley but scarce on the west side; at Death Valley IV sites on the east side, sheep bones and blinds are plentiful near springs but are infrequently found on the west side. This archaeological record suggests that the distribution of bighorn sheep has changed very little during the past thousand years, another reason not to blame the burro for restricting the range of the sheep. The archaeological record also suggests that deer have always been scarce in the Death Valley area.

Death Valley IV and possibly earlier Indians also decorated the varnished surfaces of rock faces with petroglyphs (fig. 131), mostly near springs. The pecking at these locations clearly is younger than the stain of iron and manganese oxide desert varnish (see p. 83), except at wet or moist locations, and provides some of the evidence for dating the stain on the rock. Only rarely did the Indians paint pictographs on rock surfaces.

9 Archaeology since 1849

The first white people arrived in Death Valley in 1849. They were part of a large wagon train of emigrants, led by a Mormon named Jefferson Hunt, and were headed for the California goldfields via Los Angeles. Near Las Vegas some of them decided to quit the wagon train and take a more westerly shortcut, a route that took them into Death Valley. One group, calling themselves the Jayhawkers, entered Death Valley on Christmas Day, 1849; a second party, including the families of Asabel Bennett and J. B. Arcane, arrived a few days later.

The Jayhawkers turned northwest, presumably following the trail around the north side of the salt pan. In the vicinity of McLean Spring or at the foot of Emigrant Wash they are supposed to have burned their wagons in order to smoke their meat. It is understandable that they might abandon their wagons rather than try to take them up Emigrant Wash, but why, in an area where mesquite was plentiful and easily gathered, they should burn wagons as fuel has not been explained.

The Bennett-Arcane party followed the trail south from Furnace Creek Wash, perhaps because that trail, after crossing to the west side of the salt pan, led to better water than had been found by the Jayhawkers. According to signs erected in Death Valley, the party camped at Tule Spring for about a month while two members explored the route beyond and sought assistance. Eagle Borax Spring, however, has more water, more trees, and better ground. That Eagle Borax and not Tule Spring may have been the site of their camp is suggested by the fact that the Wheeler Survey in the 1870s referred to that spring as Emigrant Spring.

While the main party rested at the spring, whether Eagle Borax or Tule, two members, William L. Manly and John Rogers, set out on foot across the Panamint Range, finding a route 250 miles across mountains and deserts to the San Fernando Valley. There they obtained food and supplies, returning after twenty-five days to the beleaguered party in Death Valley. Manly and Rogers led the party out of Death Valley on foot by the route they had taken, supposedly up Six Spring Canyon. A plainly marked route with more frequent water holes and easier grades follows the Indian trail southwest from Bennetts Well to the springs at White Tanks between the mouths of Starvation and Johnson canyons. From White Tanks, any one of several trails leads to good water and the west side. This possible route does not fit details of the journey as described by Manly, but his account was written almost a half century later. That he and his companion successfully made the trip to the San Fernando Valley and returned is evidence that they followed Indian trails that led to water and were not wandering aimlessly across intervening deserts.

OLD TRAILS

Old abandoned trails, relics of past eras, are one of Death Valley's least noticed but most interesting features (fig. 132). They were abandoned years ago when vehicles became the chief mode of travel and the old trails, suitable only for traveling by foot or horseback, fell into disuse. The date the trails were abandoned, about 1900, is recorded by the scarcity of post-1900 litter along them and by the abundance of litter of earlier times. The trails have been preserved as relics because Death Valley is not suitable for grazing; only small animals have used the trails since they were abandoned by travelers.

That the trails were used by prospectors before 1900 is clearly recorded by the litter: early-type tin cans, early-type bottles, pack equipment with square nails, heavy baling wire, and other pre-1900 articles. That the trails were used by Indians long before the white men arrived is clear because they lead to archaeological sites, and Indian artifacts are common along them. Moreover, the trails are direct routes between springs, and a main trail in such a desert is as conspicuous as a highway. There is little doubt that pioneers found their way from spring to spring by following the old trails. Mapping them led to rediscovery of the original road across the salt pan north of the Devils Golf Course (fig. 133).

Not only are the trails conspicuous, and the shortest routes between springs, but one soon learns to tell in which direction along a trail is the nearest spring. The location of water is shown

FIG. 132. Old trail across gravel fans in Death Valley, south from Furnace Creek fan where trail crosses ridge south of Mushroom Rock.

by the way the coyote trails and little-used grazing trails branch from the main one; the branches converge as a trail approaches a spring and then diverge from it.

The trails have been mapped and studied in some detail in order to estimate rates of erosion. For the most part, those on older gravel deposits and subject only to local runoff are intact. On younger gravels, subject to heavier runoff, 10 to 20 percent of the old trail alignments may have been destroyed, and the destroyed stretches invariably are in swales where the runoff is concentrated. Even on the youngest gravels, subject to present-day washing, as much as 25 percent of the alignments may be preserved. The trails add their bit to other evidence indicating very slow erosion under the present climatic regimen. Other evidence of slow erosion is provided by (1) the extent of destruction of (*a*) roads, flood-control ditches, and other features constructed since Death Valley became a national monument; (*b*) prehistoric archaeological features, such as those illustrated in figures 113, 122; (*c*) fault scarps believed to be about 2,000 years old (figs. 92,

FIG. 133. Tracks made by 20-mule teams across rough, silty rock salt formed original road across Death Valley. The road was found by following old trails and was identified by litter discarded along the way.

93); and (2) the extent of weathering, erosion, and sedimentation on Quaternary deposits, such as formations on the salt pan and on gravel fans.

THE ARCHAEOLOGY OF LITTER

Tin cans, bottles, and other litter discarded along old trails and at mining camps can be used for dating the trails, because the litter reflects changes in the technology of manufacture.

Before the Owens bottling machine was invented in 1902, bottles were made by hand. The two kinds of manufacture can be distinguished by the seam marking the two edges of the mold. In machine-made bottles the seam extends all the way to the top of the bottle and across the rim. The neck of a bottle that has been finished by hand, however, has the seam below the neck (fig. 134). The handmade product was broken from the blowing iron and the bottle maker then heated the fractured neck and finished it with a band of molten glass. In a hand-finished bottle the neck was turned, and the seam commonly ends in a twist.

Although bottles were made by machine as early as 1902, they did not become plentiful in Death Valley for another dozen years. An abundance of machine-made bottles therefore indicates a date later than about 1915; an abundance of bottles with hand-finished necks points to an earlier date.

There are several other ways in which litter helps with dating. Beer bottles and soft-drink bottles with hand-finished necks were made, before about 1900, for cork stoppers only; after that time they were made for metal caps (fig. 135). The shift from square nails to round nails also occurred about 1900, and there was a change in the weight of wire used for baling hay. Pre-1900 baling wire is about twice the diameter of wire made after that time. About the time of World War I the method of manufacturing tin cans changed. Before that time, tin cans were sealed with solder; later they were sealed by crimping the edges (fig. 136).

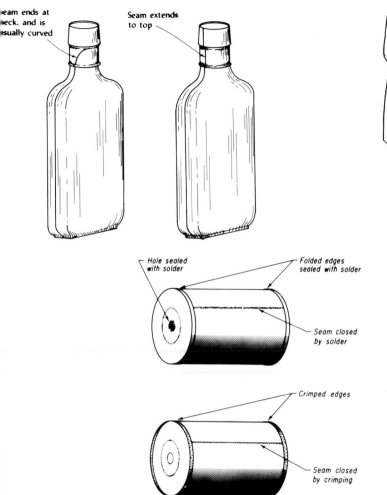

Seam ends at neck, and is usually curved

Seam extends to top

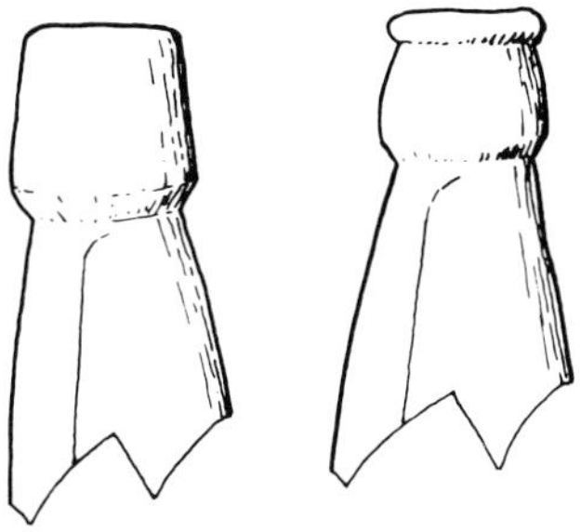

Hole sealed with solder

Folded edges sealed with solder

Seam closed by solder

Crimped edges

Seam closed by crimping

FIG. 134. Old-type bottle with hand-finished neck (left) has mold seam ending at neck, usually in a curve. On machine-finished bottle (right), seam extends to top.

FIG. 135. Neck of pre-1900 beer bottle, made for cork stopper (left); neck of post-1900 bottle, for beer or soft drinks, made for metal cap (right). Both necks here illustrated were hand finished.

FIG. 136. Old-type tin can (top); modern type, with crimped seam and ends (below).

In brief, the pre-1900 litter includes bottles with hand-finished necks made for cork stoppers. These are associated with square nails, soldered tin cans, heavy baling wire, wagon parts, and packsaddle equipment.

Litter discarded during the first two decades of the twentieth century is transitional. Probably the most satisfactory "index fossils" are bottles with hand-finished necks made for metal caps, but these are mixed with hangers-on of the older-type bottles and some pioneering machine-made bottles. The tin cans are soldered. Round nails are mixed with square nails. Wagon parts are mixed with automobile parts.

After 1920 the litter is altogether different. Tin cans are crimped, bottles are machine made, and nails are round. Instead of wagon parts and baling wire one finds magneto boxes, car tires about 3 inches diameter, and that infamous monkey wrench known as the knuckle breaker.

Litter discarded during and since the latter part of the 1930s includes all these artifacts, plus the beer cans with its familiar triangular opening, and aluminum cooking utensils. For this period, it seems likely that the beer can will become the index fossil because can openers were changed about 1950. Pre-1950 openers made a triangle 30 millimeters long; later openers make a triangle barely half that length. And now there are the self-openers. The abundance of beer cans shows that Death Valley residents during this stage were carrying on increased trade with urban centers.

Glass   The feature of old mining camp litter which commands the widest interest is purple glass. A visit to an old camp has been a success if an intact purple bottle can be retrieved, for such items are becoming scarce as the hobby of collecting gains popularity. The chemistry of glass and its coloration is a fascinating history. The subject has been investigated intensively by physical chemists, but a few of the highlights relating to utility wares of the kind found at mining camps, mostly glass bottles, are worth repeating.

Some of the coloring materials in glass are in solution in the glass and color it the same way that water becomes colored when a colored salt is dissolved in it. Other coloring materials are probably in mechanical suspension, and these particles can absorb the light rays characteristic of the color they produce.

The composition of glass varies in accordance with its intended use. In utility glass, such as bottle glass, silica is combined with lime and soda. Other glasses, used chiefly for beer, wine, and spirit bottles, also contain alumina which seems to add to strength and insolubility. Finer-quality silica glass, like Bohemian glass, is lime-potash. Some optical silica glass is lead-potash glass and is known as lead flint.

Pure silica glass would be the most suitable material for most uses, but the cost of manufacture would be prohibitive because both the melting temperature of silica and its viscosity are high. The latter quality makes it difficult to free the melt of gas bubbles. Other oxides are therefore added to lower the melting point and the viscosity.

Alkalies added to glass increase fusibility and lower viscosity. These alkali compounds usually melt first, and the fused compounds begin to react with and flux the silica. Finally the alkaline earth oxides are dissolved in the resulting alkali silicates. The fluxes most commonly used to supply the alkalies are sodium carbonate, sodium sulfate, sodium borate, sodium nitrate, potassium carbonate, and potassium nitrate. Alkalies make melting easier and cheaper, but they also render the glass more susceptible to corrosion. In general, susceptibility to corrosion increases with an increase in alkalies, especially sodium.

Bottle glass at old mining camps is likely to have its exposed surfaces corroded, for the sodium content was not well controlled until machine methods were adopted. Such glass exposed to acid vapors develops an irridescent surface because moisture containing atmospheric carbon dioxide reacts with the glass surface, forming a layer composed of hydrated silica and carbonated sodium and calcium oxides.

Bottle glass is soda-lime glass. In such glass too much soda increases the susceptibility to corrosion; too much lime leads to crystallization; too much silica requires excessive temperatures for fusion and for working. Even an ancient glass from dynastic Egypt is typical soda-lime glass, containing roughly 65 to 70 percent silica, 6 to 10 percent calcium and magnesium, and 16 to 23 percent alkali. Glass in pre-World War I bottles is like ancient glass, but most bottles made since World War I are, on the average, slightly higher in SiO_2 and slightly lower in Na_2O and CaO.

It may seem surprising that the composition of ancient and modern glass is similar, inasmuch as early factories had to use raw materials that were nearby. Early European glass factories were in forests where fuel could be obtained cheaply and where crude potassium carbonate could be obtained by lixiviation of wood ashes. Soda ash, obtained from seaweed, was used in factories located near coasts. Window glass and bottle glass had roughly the same composition.

The similarities in glass composition through the ages arise from the physical chemistry of glass, especially the phase equilibrium of soda, lime, and silica. Lowest melting temperatures are obtained when the proportions approximate 75 percent SiO_2 and 25 percent Na_2O, but lime must be substituted for part of the alkali to achieve stability against corrosion. Yet too much lime or too much silica introduces other difficulties. Although they did not understand why, the ancients learned by trial and error what range in proportions was feasible for manufacturing glass.

Returning now to the question of color, the purple glass at old mining camps originally was clear, but exposure causes photochemical changes in manganese oxides which in turn cause the glass to become purple. When glass manufacture was largely by hand, the manufacturer could adapt the process to the material at hand, but when methods became mechanized, the material had to be adapted to the process, and less variation in composition could be allowed. Apparently, since the advent of machine-made bottles, the materials used in making glass contain fewer impurities that would change the color of the glass.

That the purple color in old glass is due to exposure has been demonstrated by experiments in which glass partly covered with paint was exposed to sunlight. When the paint was removed the exposed part was colored, whereas the protected part was not. Anyone who finds glassware partly buried in the ground can satisfy himself as to the truth of the statement; the part that was buried remains clear while the part that was exposed has become purple (fig. 137). The color change is by no means peculiar to deserts and high altitudes; it occurs also in tropical and temperate regions and at low altitudes. It also has been obtained by exposing glass containing manganese to ultraviolet rays from a mercury vapor lamp. That the purple color is due to manganese oxide in the glass as well as to exposure is further confirmed by a series of spectrographic analyses of various colored glass from old mining camps in Death Valley (table 5).

The length of time required for glass to become purple depends partly on composition of the glass, especially its manganese content, partly on exposure to sunlight, and partly on the color of the background. Given optimum conditions the color change can occur in less than a month. Exposure of less than a year produced a violet color in most old glass containing appreciable

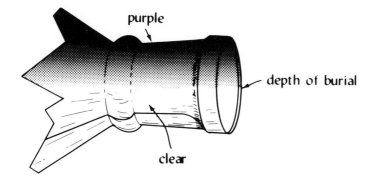

FIG. 137. Rim of partly buried old jar with hand-finished neck. Exposed part is purple; buried part has remained clear.

Table 5. PARTIAL SEMIQUANTITATIVE SPECTROGRAPHIC
ANALYSES OF OLD GLASS FROM PRE-WORLD WAR I
MINING CAMPS IN DEATH VALLEY
(In percentages)

Type of bottle	Coloration	Manganese	Iron	Arsenic
Spice and medicine	A-1 Deepest purple . .	0.70	0.05	0.15
	A-2	0.70	0.07	0.15
	A-3	0.70	0.10	<0.05
	A-4	0.20	0.05	<0.05
	A-5 Least purple	0.30	0.15	0.07
Whiskey, with cork	B-1 Deepest purple. . .	>1.00	0.10	0.20
	B-2	>1.00	0.07	0.05
	B-3	0.70	0.07	<0.05
	B-4	0.15	0.05	<0.05
	B-5 Least purple	0.002	0.03	<0.05
	L-1 Brown	0.002	0.07	
Soft drink, with metal cap	C-1 Pale blue	0.01	0.15	0.05
	C-2 Slight purple	0.10	0.10	0.15
	C-3 Pale blue	0.015	0.15	0.15
	C-4 Pale green	0.005	0.07	0.10

ANALYSTS: E. F. Cooley and U. Oda, U.S. Geological Survey.

quantities of manganese. In some bottles, the color change occurred before the gummed paper labels were destroyed. The color becomes more pronounced as the time of exposure is lengthened. Background colors seem to affect the rate of color change too. Violet-colored backgrounds accelerate color change, presumably by favoring ultraviolet rays; black and brown backgrounds seem to retard the change. Backgrounds of white, yellow, blue, and red apparently have no influence. Backgrounds containing manganese have no effect.

Modern bottles seldom become purple because their manganese content is usually low, not because the time of exposure has been insufficient. The clear glass in modern grocery bottles, and in modern wine, whiskey, and gin bottles, contains less than 0.002 percent manganese.

Metal oxides other than manganese produce colors in glass. Reds may be produced by selenium or, if oxidation is retarded, by copper. Orange is obtained with cadmium. Yellows are caused by uranium, cesium, titanium, or indium, though the yellow color of signal glass commonly is obtained by adding carbon and sulfur to a soda-lime glass. Iron may produce yellow, green, or blue glass, depending on the proportions of ferric and ferrous iron; in turn this proportion is dependent on the temperature of melting and the presence or absence of substances like arsenic, selenium, and manganese which upset the ferric-ferrous equilibrium. Manganese and iron together in quantity produce amber or brown. Chromium added by itself colors glass bright green, but mixed with iron or copper it produces blue or green. Copper produces green if oxidation is complete; green signal glass commonly contains a percentage of cupric oxide in a soda-lime glass. Cobalt produces blue. Alkalies and alkaline earths have little or no coloring effect.

10   Plant and Animal Geography

Horticultural research in Death Valley supports the principle
that, for healthy growth, plants need water. Whatever controls the
occurrence of water controls the occurrence of plants.

The first factor controlling the occurrence of water is climate,
both latitudinal and altitudinal. The second factor is geology;
within a given climatic zone, the occurrence of water for plant
growth is controlled chiefly by geologic conditions. Plants are
good geologists and good hydrologists, and many of the plants in
Death Valley are very good chemists too. In Death Valley the distribution of a species so closely accords with the geology that a
map of the one is very nearly a map of the other.

Regional factors are best shown on small-scale maps of large
regions. A plant map of a large region closely resembles a
climate map of that region. The regional distribution of species
correlates with regional differences in climate which reflect differences in latitude, longitude, and altitude, as pointed out in
1898 by C. H. Merriam.

CLIMATIC ZONING OF PLANTS

The principal plant zones recognized in western North America
are, from north to south: Alpine zone, Hudsonian zone, and

Canadian zone in the Boreal region; Transition zone; Upper Sonoran zone and Lower Sonoran zone in the Austral region.

The treeless Alpine zone and the Hudsonian zone, which is characterized on western mountains by spruce and fir, are not represented in the Panamint Range. The highest zone in Death Valley is the Canadian zone, marked by limber pine and bristle-cone pine (figs. 138, 139) on the summit north and south of Telescope Peak. Also missing is the Transition zone which, on most western mountains, is characterized by western yellow pine. The Panamint Range is one of very few ranges in the Southwest high enough to reach into the Transition zone, but yellow pine is rare.

The Upper Sonoran zone is represented by the woodland of piñon and juniper and by the zone with shadscale and other shrubs below the woodland. The Lower Sonoran zone, which is treeless except at places with water, is marked by the creosote bush. The Lower Sonoran zone covers the gravel fans and the lower mountain slopes to an altitude of about 4,000 feet.

Although the distribution of principal plant formations and floral zones is climatic, there is also a geologic factor to the extent that climates have changed drastically during Quaternary time. A study of the present distribution of species therefore needs to be put in the context of how the climatic environment has changed during Pleistocene and Holocene time. At Gypsum Cave, Nevada, northeast of Las Vegas, there is evidence from sloth dung that in latest Pleistocene time, when the cave was first occupied by man, high-altitude plants were growing 3,000 feet lower than they are now. The vegetation in Death Valley below the levels reached by the Pleistocene and Holocene lakes has moved there since the time of the lakes.

WATER SUPPLY OF PLANTS

Where groundwater is shallow enough to be within reach, plants send their roots to the water table or capillary fringe and have a

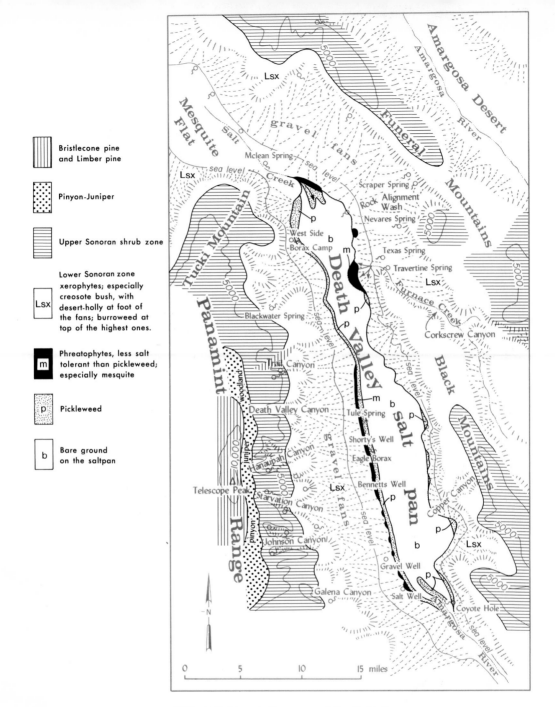

FIG. 138. Vegetation map of Death Valley.

perennial water supply (fig. 140). Such plants are known as phreatophytes. Where the water table is too deep to be reached by plant roots, there grows another type of plant—xerophytes—which can survive protracted drought. They depend on the ephemeral water (vadose or suspended water) in the ground above the water table, which resembles soil water except that it is in the gravel or other parent material below the surface layers of soil. In Death Valley the distribution of phreatophytes and xerophytes is orderly, reflecting the geologic control of groundwater occurrence.

FIG. 139. Three varieties of pine grow at different altitudes on Panamint Range. On the highest level, at Telescope Peak, is the bristlecone pine (right) also known as foxtail pine (*Pinus aristata*). It has needles in bundles of five and cones with a long bristlelike prickle. Also on the summit, but at lower altitudes, is the limber pine (*Pinus flexilis*; center). It also has needles in bundles of five, but they are differently distributed on the branchlets, and the cones have no prickles. At lower altitudes, down to about 5,500 feet, grows the piñon (*Pinus monophylla*; left), which is characterized by single needles and by small cones yielding edible seeds much prized by the Indians.

On the gravel fans sloping down from the mountains, the water table is deep and the plants are xerophytes (fig. 141). At the foot of the fans the gravel grades laterally to sand and silt, and groundwater there is ponded against the side of the fine-grained sediments. Groundwater there is shallow; there are even some springs. This belt of shallow groundwater surrounding the salt pan supports stands of phreatophytes.

The salt pan itself is without flowering plants; it is bare ground, which is conservatively defined as ground that has less than one shrub per acre. As a matter of fact, the salt pan covers about 200 square miles and has no shrubs at all. About 30 percent of the area of the gravel fans is also bare. In these areas, for geologic reasons, the supply of vadose water is deficient.

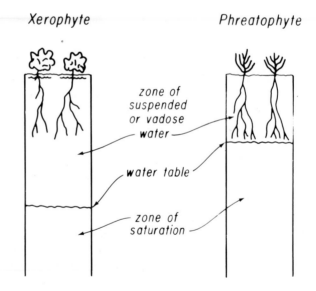

FIG. 140. Water supply of plants. Xerophytes depend on ephemeral water in ground above water table, the zone of suspended or vadose water. They are capable of surviving protracted dry periods. Phreatophytes, whose roots go down to the water table, have a perennial water supply from top of saturated zone.

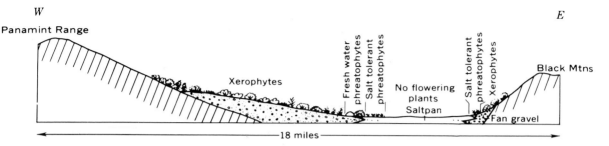

FIG. 141. Generalized transects across Death Valley showing distribution of woody plants. Xerophytes grow on gravel fans and extend into mountains. Phreatophytes grow at foot of fans, where groundwater is shallow. Fresh-water phreatophytes grow only on west side of Badwater Basin and east side of Cottonball Basin, where fans are long and high. Only salt-tolerant phreatophytes grow where salt pan is crowded against foot of adjoining mountains. There are no flowering plants on salt pan. (From U.S. Geol. Survey Prof. Paper 509.)

Xerophytes   The shrub known as desert holly (*Atriplex hymenelytra*; fig. 142) is the most drought-resistant—most xeric—plant in Death Valley. It grows on the hottest, driest, and saltiest parts of the gravel fans where the ground is too dry or too salty even for creosote bush, the characteristic plant of the Lower Sonoran zone which is the most extensive grower on the gravel fans and lower parts of the mountains.

Desert holly forms nearly pure stands at the foot of fans around the northern part of the salt pan and along the east side (fig. 143). Most of the fans contain a high percentage of carbonate rocks and all of them are saline. Plant densities range roughly from 5 to 250 shrubs per acre. The lower boundary of the desert holly is the edge of the salt pan; where there are phreatophytes at the edge of the pan there is surprisingly little mixing. In places where gravels have been washed onto the pan the desert holly extends panward along the gravelly wash.

The only other perennial shrubs that commonly grow in the nearly pure stands of desert holly are honeysweet (*Tidestromia oblongifolia*) and a species of spurge (genus *Euphorbia*). These shrubs grow extensively along the east side of Death Valley, north of Badwater. Locally there is sparse inkweed (*Suaeda suffrutescens*) growing with desert holly on the lowest parts of the fans, especially where the ground is saline.

Desert holly is more abundant on the east side of Death Valley than on the west side, probably because the fans on the east side are saline as well as dry, and desert holly thrives on salt. In the area extending south from Badwater to the south side of Copper Canyon fan, where fan gravels are notably saline, desert holly is practically the only shrub that flourishes.

Even this salt-loving plant has its limits. In parts of the gravel fans where there is excessive runoff, desert holly cannot live. In other places it may be able to grow along washes where runoff is collected from hillsides but not be able to move onto those hillsides. If there is a succession of wet years, however, desert holly

FIG. 142. Desert holly (*Atriplex hymenelytra*).

may be able to move into rills on the hillsides, though it would
be unable to maintain itself there through a later stretch of dry
years; in such a period of aridity the hillside rills are lined with
dead desert holly. In other places the ground may be too salty for
desert holly. The shrub averages about 30 to 35 percent ash
which is half sodium chloride, but it does not grow on ground
containing more than roughly 2 percent (by volume) of water-
soluble salts (fig. 144).

The composition of the ash in Death Valley desert holly helps
to identify the source of water utilized by xerophytes. Most of the
water must come from the vadose zone in the ground and not
from the atmosphere as dew, because the chemistry of the plant
ash correlates with the chemistry of the ground. If molybdenum
is plentiful in the ground, as it is in the vicinity of Cow Creek,
the plant ash contains more than the average amount of that

FIG. 143. Desert holly forms nearly pure stands along edge where gravel
fans border salt pan. Stand shown here is 0.5 mile northwest of Park
Service service area.

element. Where the ratio of calcium to sodium is the highest, the ash of desert holly contains 20 percent more calcium than it does where the ratio is low.

In the southern part of Death Valley and in a small area north of Salt Creek, nearly pure stands of cattle spinach (*Atriplex polycarpa*; fig. 145) grow on the lower parts of the gravel fans, below the creosote bush (fig. 146), a position that in the rest of Death Valley is occupied by desert holly. Cattle spinach grows on nonsaline fans which do not have high percentages of carbonate rocks; these fans are derived chiefly from Precambrian and Lower Cambrian formations along the west side of the salt pan. Mounds of sand a foot high have accumulated around the shrubs.

Some nearly pure stands of cheesebush (*Hymenoclea salsola*) grow along washes at the foot of some fans on the west side of Death Valley. Around the salt pan the shrub is restricted to these washes, but elsewhere in the region it grows at higher altitudes.

The next most xeric shrub after desert holly, in this area, is creosote bush (*Larrea tridentata*; fig. 147). It extends from near the foot of the fans (fig. 148), about 240 feet below sea level, to an altitude of about 4,000 feet in the mountains. Creosote bush, along with desert holly and cattle spinach, covers most of the lower half or lower two-thirds of the gravel fans up to about 500 feet above sea level.

Four floras dominated by creosote bush may be distinguished. At the lower edge of the northern fans, where creosote bush adjoins stands of desert holly, the two shrubs are mixed. On fans farther south, where creosote bush adjoins stands of cattle spinach, these two shrubs are mixed, but somewhat higher on the same fans this combination is replaced by another mixture of creosote bush and desert holly. A third flora, consisting of nearly pure stands of creosote bush, extends up the fans to the lower limit of the burroweed (*Franseria dumosa*) or incienso (*Encelia farinosa*). A fourth flora, which occurs along the washes, consists of creosote bush with or without desert holly and cattle spinach but mixed with such shrubs as desert trumpet (*Eriogonum*

FIG. 144. Some gravel fans, especially leeward from Cottonball Basin, have catchment areas and infiltration rates that would favor the growth of desert holly, but the ground is saline because of windblown salts. Desert holly here, north of highway above North Side Borax Camp, grows in the wash; the low bench to the left is bare, saliferous gravel, and even the stones there are disintegrating.

FIG. 145. Cattle spinach (*Atriplex polycarpa*).

inflatum), stingbush (*Eucnide urens*), sticky-ring (*Boerhaavia annulata*), and honeysweet.

The density of creosote bush ranges from very few to about 125 plants per acre; the average probably is near 40. Creosote bush has less ash than desert holly—less than 10 percent—and very little of it is sodium chloride.

The distribution and healthiness of creosote bush are in proportion to the suitability of the ground for recharge of vadose water. Along the sides of paved roads, for example, creosote bush is much healthier than it is away from the roads because of the slight additional runoff from the pavement. It can grow along washes that collect runoff from bordering hillsides that are bare. In sandy ground it can even survive a good deal of wind erosion, which exposes its roots.

FIG. 146. Xerophyte plants on lower part of Johnson Canyon fan. Beyond cattle spinach in foreground is dark creosote bush. Farther up, gravels are darkly stained with desert varnish on surface of desert pavement, which favors runoff and minimizes infiltration; the ground is bare. Gravels here and farther south in Death Valley are mostly derived from Precambrian formations, and the ground is sandier than it is farther north. Cattle spinach grows on toes of sandy fans; desert holly grows on toes of less sandy fans.

FIG. 147. Creosote bush (*Larrea tridentata*) is characteristic of Lower Sonoran zone.

FIG. 148. In this typical stand of creosote bush on fan just west of Shorty's Well there are about 150 shrubs per acre, but individual plants are small and seem to be struggling. A few desert holly plants grow with the creosote bush.

The stands of creosote bush are most dense where surface layers favor seepage into the ground and where a catchment area augments the supply of surface water discharging onto that ground. Such environments are provided by the youngest gravels; at only a few places does creosote bush extend onto the oldest gravel surfaces, where the desert pavement favors runoff. Also, bare pavement areas are virtually without annuals except in very wet years.

Healthy stands of creosote grow on the fans tributary to Cotton-ball Basin, but southward along the east side of Death Valley, from the middle of Cow Creek fan to the south side of Copper Canyon fan, there is no creosote bush except for a few individuals along the irrigation ditch at Furnace Creek Ranch and half a dozen scattered bushes on the fans south of Badwater. The absence of creosote bush probably is attributable to the high rate of runoff on the fans of fine-grained sediments north of Badwater and the high salinity of fan gravels south of Badwater. The dryness on these fans is accented by their exposure: they not only face west, but they are at the foot of a high rocky scarp that must reflect a great deal of heat from the afternoon sun.

Least xeric of the xerophytes are burroweed (fig. 149) and incienso, which grow on the highest parts of the fans, usually about 500 feet, and extend upward into the mountains. Burroweed is found in the northern part of the valley; its place in the south is taken by incienso. In general, burroweed occurs on fans that have pure stands of desert holly at the base, whereas incienso grows on fans that have stands of cattle spinach at the base. The occurrence of these two plants may reflect the composition of the gravels: fans at the north are derived in large part from Paleozoic limestone formations (see chap. 4), whereas the more sandy southern fans are derived in large part from Cambrian and Precambrian rocks containing little limestone.

The stands of burroweed or incienso along the bottoms of washes commonly are flanked by rows of a more drought-resistant xerophyte, desert holly or creosote bush (fig. 150). The high parts

FIG. 149. Burroweed (*Franseria dumosa*).

FIG. 150. Stand of burroweed near road along Furnace Creek Wash above Corkscrew Canyon. Typically this shrub grows high up on fans, above main stands of creosote bush. Burroweed grows at bottom of washes between bare surfaces of fans with desert pavement, whereas desert holly flourishes along sides of washes.

of the fans have the largest area of old gravels with desert pavement surface; they also have the heaviest runoff and the most bare ground. Burroweed and incienso growing in the washes benefit from the runoff and form the densest stands of xerophytes on the fans; they may be spaced so closely that their crowns touch.

A xerophyte of more than passing interest is beavertail cactus (*Opuntia basilaris*), which the Indians used. It grows mostly on the fans north and east of Cottonball Basin. Another xerophyte used by the Indians is the gourd (*Cucurbita palmata*), which grows in canyons at altitudes above 1,000 feet.

The distribution of annuals in Death Valley accords with the distribution of xerophytes. A study by Fritz Went and his colleagues shows that Death Valley has a good crop of annuals about once every six or seven years when autumn and winter rainfall is twice the average for those seasons. At that time the valley becomes carpeted with flowers. Annuals are most abundant where the supply of vadose water is the largest, that is, in stands of burroweed. They are next most abundant where creosote bush flourishes, and next in the areas of desert holly. In very good years annuals grow on parts of the fans which are bare of perennials. Even in favorable years, few annuals grow in the belt of phreatophytes at the edge of the salt pan, and none grow on the salt pan.

Phreatophytes   Phreatophytes are plants whose roots reach perennial groundwater or the capillary fringe above the water table. Nine varieties are found around the edge of the salt pan; others grow at springs on the gravel fans and at other places where the water table is shallow. The distribution of phreatophytes is controlled not only by the availability of groundwater but also by its quality. Arranged in order of increasing tolerance to salinity of the groundwater and by locality, the principal phreatophytes in Death Valley area are:

(at springs on gravel fans)
    desert baccharis (*Baccharis sergiloides*)
    willow (*Salix* spp.)
    screw-bean mesquite (*Prosopis pubescens*)
    common reed grass (*Phragmites communis*);
(around the edge of the salt pan—nine varieties)
    honey mesquite (*Prosopis juliflora*)
    arrowweed (*Pluchea sericea*)
    four-wing saltbush (*Atriplex canescens*)
    alkali sacaton grass (*Sporobolus airoides*)
    tamarisk (*Tamarix gallica; T. aphylla*)
    inkweed (*Suaeda* spp.)
    salt grass (*Distichlis stricta*)
    rush (*Juncus cooperi*)
    pickleweed (*Allenrolfea occidentalis*)

Salinity of the groundwater under phreatophytes ranges from a few hundred parts per million under mesquite, the least salt tolerant, to a maximum of about 6 percent under pickleweed. The zoning of phreatophytes is illustrated by the transects in figure 151.

Pickleweed, the most salt-tolerant plant, grows in large pure stands extending from the bare salt pan to the zone where arrowweed is found. The latter, usually including some pickleweed, extends into the mesquite belt where the two plants are generally mixed. In Death Valley no annuals grow on highly saline ground where pickleweed or salt grass is found, and few if any grow where there is arrowweed. Pickleweed readily seeds itself. It grows in furrows of old borax workings and even in ditches bordering the present highway.

The stands of pickleweed form a belt ranging from a quarter mile to 2 miles wide along the whole west side of the salt pan; along the east side there are only isolated patches. Pickleweed is a curious sprawling succulent shrub. Individual plants are usually less than a foot high, but they may spread over several square feet. Depending on the season and on the availability of water, the color of pickleweed is grass green, yellow, brown, or red. Between mounds of pickleweed the ground is covered with a blisterlike crust of silty salt. The water in the ground around pickleweed roots contains up to 6 percent salt, thus having almost twice the salinity of seawater. Total ash in pickleweed tops is as high as 40 percent dry weight; the ash is mostly sodium chloride. Yet it is possible to recognize plants that are alive.

Two species of tamarisk, the so-called salt cedars, grow in Death Valley. Although they have been introduced into the area, they are well suited to the environment and could spread sufficiently to become a pest. Tamarisk seems to do equally well in the pickleweed zone and where the water is less salty.

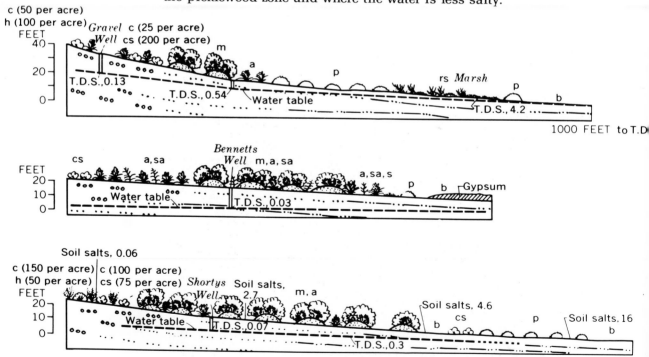

FIG. 151. Transects showing relationship of phreatophytes to quality of water on west side of Death Valley, at Gravel Well, Bennetts Well, and Shorty's Well. *c.* Creosote bush. *h.* Desert holly. *cs.* Cattle spinach. *m.* Honey mesquite. *a.* Arrowweed. *sa.* Sacaton grass. *s.* Salt grass. *r.* Rush. *p.* Pickleweed. *b.* Bare ground. *T.D.S.* Total dissolved solids in groundwater, in percentages. Water-soluble soil salts are given in percentages by volume. (From U.S. Geol. Survey Prof. Paper 509.)

Salt grass (fig. 152) and rush grow together where the ground-water contains 3 or 4 percent salts. Rush has roots that extend downward, whereas salt grass is characterized by spreading laterals (runners or rhizomes). As a result, clumps of salt grass tend to be arranged linearly. Inkweed (*Suaeda suffrutescens*) occurs in mixed stands where the salinity is somewhat higher than at the stands of salt grass and rush; the groundwater under stands of inkweed may contain 5 percent salts.

Arrowweed (fig. 153) and four-wing saltbush (fig. 154) grow on the salt-pan side of the honey mesquite and between the mesquite and the pickleweed. Both the ground and the groundwater are more saline than under mesquite and less so than under pickle-weed. In stands of arrowweed the ground is saline. Four-wing saltbush grows where the ground is sandy and not so saline. Stands of arrowweed are open; the crowns rarely touch. A good stand includes seventy-five to a hundred shrubs per acre. Salt

FIG. 152. Salt grass (*Distichlis stricta*) is common near edge of salt pan where water in ground is no more than 3 percent salts.

grass is common with both arrowweed and four-wing saltbush. Pickleweed may also grow with them.

The chemistry of arrowweed is of some interest. It has a moderate ash content, generally 10 to 15 percent, mostly as sulfates. There is very little chloride. Moreover, sodium exceeds calcium in specimens from the east side of Cottonball Basin, an area of sodium sulfate, whereas calcium exceeds sodium on the west side of Badwater Basin, an area of calcium sulfate. Arrowweed

FIG. 153. Stand of arrowweed near highway 190 at Devils Cornfield.

accumulates sulfates but seems indifferent as to whether the salt is the readily soluble sodium sulfate or the less soluble calcium sulfate.

Honey mesquite (fig. 155) grows at the edge of the salt pan where the groundwater is shallow and not too saline (less than 0.5 percent). Since the water is potable, mesquite marks the places around the salt pan where one can obtain a drink by digging down to the water table. Foxes already know this valuable

FIG. 154. About 2 miles southeast of Salt Well, at southwest edge of salt pan, where ground is sandy, four-wing saltbush grows in washes down to edge of pickleweed stand, just beyond figure.

sign, and the desert traveler soon recognizes it. Mesquite areas, where Indians camped, possess by far the most varied fauna at lower altitudes in Death Valley.

Mesquite roots can grow to an enormous length; some have measured more than 50 feet. The plant flourishes in several environments around the salt pan, but mostly it prefers old stabilized dunes 10 feet high. Opposite Telescope Peak, where maximum fresh groundwater is entering Death Valley from the west, the water is in sandy and silty beds beneath the dunes. On the east side of Cottonball Basin the water rests on playa beds underlying gravels and sand at the toe of the fans. At Tule Spring the trees are on silty ground having a salt crust, but they are lined up as if their roots were in buried gravel or sand, either of which serves as a channel-like aquifer containing water without much salt.

In a classic report on the botany of Death Valley (1893), F. V. Coville wrote of the zoning of phreatophytes: "First there occurs a strip, a few meters broad, of *Allenrolfea occidentalis*; next a similar strip of *Juncus cooperi*; and third a belt of *Sporobolus airoides* and *Pluchea sericea* about 300 meters broad. Across the second and third belts *Distichlis spicata* occurs sparingly; . . . the fourth belt consists of *Prosopis juliflora* and *Atriplex canescens*, intermixed with *Suaeda suffrutescens*. The next belt is made up principally of *Atriplex polycarpa*, with a few scattered specimens of *Larrea tridentata*. The sixth belt is that of *Larrea tridentata*" (fig. 155).

The root systems of phreatophytes are orderly, too. In the arrowweed zone, for example, because of evaporation at the surface, the salinity of the water in the ground and in the capillary fringe increases upward, and the salts are deposited in orderly layers with chlorides at the top underlain by a layer with sulfate nodules. Arrowweed roots spread in the sulfate or lower layers, but pickleweed growing with the arrowweed has roots spreading in the near-surface, chloride-rich layers. In other words, these species are not competitive; any competition that exists is between individuals of the same species.

FIG. 155. View west at Tule Spring, looking toward Hanaupah Canyon and Telescope Peak. In foreground pickleweed grows on salt-crusted silt where ground contains about 15 percent soil salts. Shrubs in middle distance are arrowweed; soil salts there are about 10 percent and groundwater contains only 0.2 percent salts. Beyond arrowweed is mesquite on ground containing less than 1 percent salts. On fans beyond mesquite is mixed growth of desert holly and creosote bush, about fifty of each per acre.

In addition to the zoning of these flowering plants, it has been found that algae, fungi, and bacteria are similarly zoned. Algae extend farther onto the salt pan than does the pickleweed; fungi extend beyond the algae to about the 10 or 12 percent brine line. Ordinary bacteria extend to about the 16 percent brine line, and probably specialized forms exist in the most concentrated brines as they do in Great Salt Lake.

CHANGES IN PLANT STANDS

The belt of phreatophytes around the salt pan in Death Valley must have established itself within the past 2,000 years, because about A.D. 1 the ground was flooded by a shallow salty lake which gave rise to the salt pan. The dating is archaeological. Sand dunes, formed since the lake disappeared, support stands of mesquite and contain the remains of later Indian occupations characterized by the bow and arrow and by pottery (see chap. 8). But there is evidence also that the stands of mesquite have deteriorated during the past few hundred years (figs. 21, 156). In places the decline may be attributed to the cutting of wood and other human activities, but in part it seems to be due to increasing salinity of the groundwater, probably reflecting the gradual drying of the climate and therefore a gradual decrease in recharge of groundwater into the salt pan. Such drying has occurred elsewhere in the Southwest. At most places in Death Valley mesquite bushes are dying and retreating, and their locations are being invaded by more salt-tolerant species.

How can a phreatophyte seed itself? How could the seed know there is groundwater below the surface? One theory is that phreatophytes can seed themselves only in wet periods when there is enough water seeping through the vadose zone for long enough periods of time to permit the roots to follow that water

FIG. 156. Relic mounds of sand where dead mesquite bushes have been overrun by cattle spinach, about midway between Gravel Well and road to Johnson Canyon. Pickleweed and bare salt pan can be seen in distance.

downward to the capillary fringe and the water table. If there is merit in this theory, there have not been enough wet periods in the past few hundred years in Death Valley to permit new growth of mesquite.

BURROS AND BIGHORNS

The biggest and gamest animal in Death Valley is the burro fig. 157), derogatorily referred to by the Park Service as "feral." In fact, the burro is a half century less feral than the Park Service. Abandoned by prospectors who brought them to Death Valley, these animals stayed on after the prospectors left. The burro is one of the few animals that have thrived in Death Valley. Because the burro has thrived so well, Death Valley has been faced with its own population explosion.

Biologists have accused the burro of fouling water holes by trampling and by defecation and therefore of driving bighorn sheep off some of the range. The burro has even been accused of attacking bighorn sheep. But these accusations were found invalid in a study by Ralph and Florence Welles, and my own observations strongly support their conclusions.

In the first place, the range used by burros is mostly on Precambrian and Cambrian formations and on gravels derived from them; bighorn sheep prefer later Paleozoic formations. The difference is not attributable to a narrow interest in just one kind of geology; rather, the younger Paleozoic formations are largely carbonate rocks which weather with jagged edges, and the sharp edges cut the burro's hoofs. Bighorn sheep prefer the rough, craggy ground of the carbonate formations, possibly because there they are safer from predators.

It cannot be argued that the burro has driven the sheep off Precambrian and Cambrian formations on the west side of the valley, because the archaeological record indicates that sheep never did use that range very much. Indians hunted the sheep when they came to springs for water. Hunting blinds erected by Indians are common at springs in or near carbonate rocks, but there are almost none at springs in Precambrian and Cambrian formations. Moreover, sheep bones are common at prehistoric Indian sites in or near carbonate rocks but are scarce where the rocks are not carbonate, as in Precambrian and Cambrian formations. The evidence seems quite clear that the burro occupies range that has been little used by sheep during the past 2,000 years; the two animals have never seriously competed for range in Death Valley.

There may be too many burros in Death Valley, threatening to change rangeland to desert (Welles and Welles, 1961:178): "For

FIG. 157. Burros are thoroughly adapted to poor grazing land in Death Valley.

the sake of the entire biota, the burros must be controlled and their numbers kept down, . . . though not because the burro is running the bighorn off the range or destroying the water supplies, for it is doing neither. An intensive program of live capture of burros to be used as pets and pack animals has been inaugurated and promises well as a control method."

The bighorn sheep is the pampered pet of the Park Service, but I confess a partiality for the more service-minded burro. Burros build trails in the mountains, where geologists need them. Trails left by bighorn sheep are likely to be dead-end routes, leading to the rim of a cliff. The rim is not the end for the sheep, whose trail continues off the base of the cliff, but it is the end of the trail for the geologist who now must backtrack.

OTHER BIG GAME

Other game animals in the Death Valley region include deer, mountain lion, bobcat, coyote, and fox. Only the coyote and the fox are common. In many deserts of the western United States the coyote has been nearly exterminated by federal and state predators who have created "the silent night"; in Death Valley one still can hear the coyote howl at night. It is said that they howl because they ate dates at Furnace Creek Ranch and swallowed the seeds.

Coyotes cross the Death Valley salt pan (fig. 158), and so far as I know they are the only animals to do so. Coyote trails cross the salt pan below Furnace Creek Ranch and enter the shrub zone on the west side. Their tracks were the only ones seen on the salt pan while it was being mapped, a process that involved crossing the valley floor along approximately a hundred traverses at half-mile intervals.

More appealing than the coyote is the kit fox, a little fellow, curious and friendly, and hardly larger than a tomcat. Indeed, so

FIG. 158. Salt Creek opposite Furnace Creek fan is usually dry enough to be forded by four-wheel-drive vehicles. Occasionally it discharges floods large enough to erode the channel. Note coyote tracks preserved beneath water (lower right).

FIG. 159. The kit fox, a trustworthy, loyal, friendly animal.

friendly is the animal that it may join the circle around a camp-fire. The one shown in figure 159 visited us, and he was so unafraid we were sure we could have fed him by hand had we stayed at that camp more than a few nights. On a later visit he brought his mate, but she was shy and stayed at the edge of the lighted area. Communication between us and the male was complete. Bedtime would come, and the lights would go out. Each morning, in the dish, was a thank-you note consisting of a single dropping. One night we had nothing to offer and went to bed without leaving any food; the next morning, on the doorstep of the trailer, we found the usual thank-you note, and the message came clear.

SMALL GAME

The mesquite-covered dunes also abound with small game, such as rabbits, rodents, lizards, and birds. All were hunted by Indians both before and after the forty-niners were in Death Valley. Small animals considered choice were desert wood rat or

FIG. 160. Merriam kangaroo rat (*Dipodomys merriami*) visiting camp high on fan at Hanaupah Canyon. This species, smallest of the kangaroo rats, has only four toes on each hind foot; it inhabits gravel fans. Another species with four toes (*D. deserti*) inhabits dunes at foot of fans. A five-toed species (*D. panamintus*) lives in piñon woodland higher up on mountains.

pack rat (*Neotoma lepida*), kangaroo rat (*Dipodomys* spp.; fig. 160), white-footed mouse (*Peromyscus eremicus*), antelope ground squirrel (genus *Ammosperophilus*), round-tailed ground squirrel (*Citellus tereticaudus eremonomus*), jackrabbit (*Lepus* spp.), and cottontail rabbit (genus *Sylvilagus*).

Since rats and mice living in the mesquite are vegetarians, their meat should be as tasty as beef (but the steaks would be smaller); indeed, they were sought as food by the Indians (see p. 173). These rodents live on gravel fans, too, but the population there is smaller than in the mesquite. Few animals range into arrowweed and pickleweed zones panward of the mesquite.

REPTILES

Among the reptiles in Death Valley are desert terrapin (*Testudo* spp.), chuckwalla (*Sauromalus ater*), banded gecko (*Coleonyx brevis*), desert iguana (*Dipsosaurus dorsalis*), collared lizard

(*Crotaphytus collaris*), leopard lizard (*C. wislizenii*), zebra-tailed lizard (*Callisaurus ventralis*), swift (*Uta stansburiana*), horned lizard (*Phrynosoma platyrhinos*), striped lizard (*Cnemidophorus* spp.), and sidewinder rattlesnake (*Crotalus cerastes*).

FISH

Visitors are likely to be surprised when they learn that the Death Valley desert has fish. They are small fellows, 1 to 3 inches long, commonly referred to as minnows but more properly known as cyprinodonts, or pupfish. Three species, all related to Colorado River species, live in springs along the Amargosa River, and one lives below sea level in the marsh at the west side of Cottonball Basin and in Salt Creek. They are descendants of fish that became isolated in springs when the drainage that formerly connected Death Valley and the Colorado River was destroyed by the downfaulting of Death Valley and of other structural basins to the southeast (pp. 132-133), perhaps in the middle Pleistocene.

The fish contribute to the interpretation of structural and drainage history. One question relates to whether the lakes to the west which were fed by the Owens River discharged into Death Valley. The Owens Valley has two genera of desert fish, *Siphateles* and *Catostomus*, said to be related to forms in the lake basins of western Nevada. *Siphateles* also occurs along the Mojave River, but neither genus has been reported in the Death Valley–Amargosa River drainage system. Further, *Cyprinodon radiosus*, which occurs in the Owens River, is said to be more closely related to Colorado River species than are any of the three cyprinodonts living in the Amargosa drainage system. This distribution suggests that drainage from the Owens Valley to the Mojave River bypassed Death Valley.

The fish are adapted to very salty water. Some of the pools in Cottonball Marsh which are occupied by fish are saltier than seawater. The pools appear between salt crusts encasing algae, the dark matter in figure 161 (cf. fig. 19), and are connected by channels 1 to 2 feet deep through the crust and by tunnels under it.

BIRDS AND PESTS

Death Valley has a considerable bird population, including visiting ducks. Most conspicuous are big black ravens.

Pests are also prevalent in Death Valley. At times both crawling and flying insects are present in sufficient numbers to be annoying. Scorpions are common on shaded and damp ground under large stones; visitors should heed the warning, "Don't pick up rocks in Death Valley."

FIG. 161. Small minnowlike fish living in Cottonball Marsh move from one pool to another by wiggling through wet mud on low divides; the three shown here (marked by pencils) may have overextended themselves in their effort to be amphibious.

Acknowledgments

Neither the original study of Death Valley nor this summary of technical reports would have been possible without the guidance and assistance of many specialists who contributed to this project. A principal contributor was Thomas S. Lovering, who with his colleagues in the Geochemical Exploration Unit of the Geological Survey—Hubert W. Lakin, J. Howard McCarthy, F. W. Ward, Walter A. Bowles, E. F. Cooley, Albert P. Marranzino, Uteana Oda, Tennyson Myers, and H. M. Nakagawa—provided the basis for most of the chemical and mineralogical work that was done. They not only peformed the necessary analytical work in the laboratory but also visited the project to provide guidance in the field.

Don R. Mabey and Gordon W. Green contributed the results of gravity and magnetic surveys and measured the tiltmeters that were installed with the advice and guidance of Jerry P. Eaton. Charles S. Denny, author of a companion study on the Amargosa Desert, frequently visited the project and aided me in understanding the gravels and the geomorphology. T. W. Robinson guided the hydrologic studies and was joint author of the section on hydrology in Professional Paper 494-B. A. L. Washburn was prime mover and joint author in the study of patterned ground in the same paper. The participation by L. W. Durrell, of Colorado State University, made possible the joint report on the distribution of fungi and algae in Professional Paper 509. Thomas W. Stern and Ralph L. Erickson provided the needed expertise for radiometric determinations. The geological studies were first directed by James Gilluly and later by J. Fred Smith of the Geological Survey; their administrative support was essential for the breadth and the extent of those studies.

The archaeological survey by Alice Hunt was directed by William J. Wallace, then of the Department of Anthropology, University of Southern California. Dr. Wallace, under contract with the Park Service, appointed Mrs. Hunt to his staff and graciously assigned her the area where I was conducting my geological studies. The result was a team effort, with the geological information aiding the archaeological work and vice versa, as brought out even in this summary account.

The many members of the National Park Service in Death Valley gave cordial encouragement and assistance during the fieldwork. Ralph and Buddy Welles, who conducted a survey of bighorn sheep while we were in Death Valley, discovered and brought to Alice Hunt's attention numerous archaeological sites, including one of the most significant. Also, Buddy Welles assisted with some of the excavations.

Almost all the linecuts are from Geological Survey professional papers; a few from other sources are specifically acknowledged in captions. Most of the photographs were taken by John R. Stacy, artist and photographer, who visited the project specifically to prepare drawings for the professional papers. The following illustrations (including linecuts and photographs) are by Stacy: figures 2, 5, 12, 13, 15, 16, 17, 18, 19, 24, 25, 26, 29, 30, 31, 32, 33, 36, 37, 38, 41, 47, 48, 52, 53, 56, 75, 76, 82, 83, 85, 91, 92, 94, 95, 96, 98, 121, 132, 142, 143, 145, 146, 147, 148, 149, 150, 152, 153, 154, 156.

Harold E. Malde of the Geological Survey took the photographs shown in figures 6, 21, 23, 133, 144.

Bibliography

Abrams, Leroy, and R. S. Ferris. 1940, 1944, 1951, 1960. Illustrated flora of the Pacific states, Washington, Oregon, and California. 4 vols. Stanford: Stanford University Press.

American Guide Series. 1939. Death Valley. Boston. 75 pp.

Antevs, E. 1952. Climatic history and the antiquity of man in California. University of California Archaeological ser., no. 16, pp. 23-31.

Bailey, G. E. 1902. The saline deposits of California. Calif. State Min. Bur. Bull. 24. 216 pp.

Bateman, P. C., and others. 1963. The Sierra Nevada batholith, a synthesis of recent work across the central part. U.S. Geol. Survey Prof. Paper 414-D. 46 pp.

Berg, J. W., Jr., and others. 1960. Seismic investigations of crustal structure in the eastern part of the Basin and Range Province. Seismol. Soc. Am. Bull., 50:511-535.

Bird, J. B. 1967. The physiography of arctic Canada. Baltimore: Johns Hopkins Press. Esp. pp. 185-198.

Blackwelder, E. 1925. Exfoliation as a phase of rock weathering. J. Geol., 33(8):793-806.

———. 1928. Mudflow as a geologic agent in semiarid mountains. Geol. Soc. Am. Bull., 39(2):465-483.

———. 1933. Lake Manly, an extinct lake of Death Valley. Geog. Rev., 23(3):464-471.

Bobek, Hans. 1959. Features and formations of the Great Kawir and Masilehx of central Iran. Arid Zone Research Centre, University of Tehran, Publ. 2. 63 pp.

Briggs, L. I. 1958. Evaporite facies. J. Sedimentary Petrol., 28:46-56.

Browne, J. R., and J. W. Taylor. 1867. Report upon the mineral resources of the United States. Washington, D.C. 368 pp.

Bryan, K. 1919. Classification of springs. J. Geol., 27:522-561.

Carder, D. S., and L. F. Bailey. 1958. Seismic wave travel times from nuclear explosions. Seismol. Soc. Am. Bull., 48(4):377-398.

Carpenter, E. 1915. Ground water in southeastern Nevada. U.S. Geol. Survey Water-Supply Paper 365. 86 pp.

Caruthers, William. 1951. Loafing along Death Valley trails. Palm Desert, Calif.: Desert Magazine Press. 186 pp.

Chalfant, W. A. 1933. The story of Inyo. Bishop, Calif. 430 pp.

_____. 1953. Death Valley: the facts. Stanford: Stanford University Press. 160 pp.

Chatard, T. M. 1890. Natural soda: its occurrence and utilization. *In* F. W. Clarke, Report of work done in the division of chemistry and physics, mainly during the fiscal year 1887-1888. U.S. Geol. Survey Bull. 60, pp. 27-101.

Clarke, F. W. 1924. The data of geochemistry. U.S. Geol. Survey Bull. 770. 5th ed. 841 pp.

Clements, T., and L. Clements. 1953. Evidence of Pleistocene man in Death Valley. Geol. Soc. Am. Bull., 64:1189-1204.

Cornwall, H. R., and F. J. Kleinhampl. 1962. Geology of the Bare Mountain quadrangle, Nevada. U.S. Geol. Survey Geol. Quad. Map CQ-157.

Coville, F. V. 1892. The Panamint Indians of California. Am. Anthropologist, o.s., 5(4):351-362.

_____. 1893. Botany of the Death Valley Expedition. U.S. Natl. Herbarium Contrib., vol. 4. 363 pp.

Curry, D. H. 1954. Turtlebacks in the central Black Mountains, Death Valley, California. *In* R. H. Jahns, ed., Geology of southern California. Calif. Div. Mines Bull. 170, pp. 53-59.

Denny, C.S. 1965. Alluvial fans in the Death Valley region, California and Nevada. U.S. Geol. Survey Prof. Paper 466. 62 pp.

Denny, C. S., and H. Drewes. 1965. Geology of the Ash Meadows quadrangle, Nevada-California. U.S. Geol. Survey Bull. 1181-L, pp. L1-L56.

Diment, W. H., and others. 1961. Crustal structure from the Nevada test site to Kingman, Arizona, from seismic and gravity observations. J. Geophys. Research, 66:201-214.

Drewes, H. 1963. Geology of the Funeral Peak quadrangle, California, on the east flank of Death Valley. U.S. Geol. Survey Prof. Paper 413. 78 pp.

Durrell, L. W. 1962. Algae of Death Valley. Am. Microscop. Soc. Trans., 81:267-273.

Dutcher, B. H. 1893. Piñon gathering among the Panamint Indians. Am. Anthropologist, o.s., 6(4):377-380.

Eaton, J. P. 1959. A portable water-tube tiltmeter. Seismol. Soc. Am. Bull., 49:301-316.

Engel, C. G., and R. P. Sharp. 1958. Chemical data on desert varnish (California). Geol. Soc. Am. Bull., 69:487-518.

Evernden, J. F., and R. W. Kistler. 1970. Chronology of emplacement of Mesozoic batholithic complexes in California and western Nevada. U.S. Geol. Survey Prof. Paper 623. 42 pp.

Ferris, R. S., and J. R. Janish. 1962. Death Valley wildflowers. Death Valley: Death Valley Natural History Association. 141 pp.

Feth, J. H. 1961. A new map of western coterminous United States showing the maximum known or inferred extent of Pleistocene lakes. U.S. Geol. Survey Prof. Paper 424-B, pp. B110-B112.

Gale, H. S. 1914. Notes on the Quaternary lakes of the Great Basin, with special reference to the deposition of potash and other salines. U.S. Geol. Survey Bull. 540, pp. 339-406.

Glasscock, C. B. 1940. Here's Death Valley. Chicago: Grosset and Dunlap. 329 pp.

Grabau, A. W. 1920. Geology of the nonmetallic mineral deposits other than silicates. Vol. I: Principles of salt deposition. New York: McGraw-Hill. 435 pp.

Grinnell, Joseph. 1937. The mammals of Death Valley. Proc. Calif. Acad. Sci., 4th ser., 23(9):115-169.

Hall, W. E., and E. M. MacKevett. 1958. Economic geology of the Darwin quadrangle, Inyo County, California. Calif. Div. Mines Spec. Rept. 51. 71 pp.

Hamilton, W., and W. B. Myers. 1967. The nature of batholiths. U.S. Geol. Survey Prof. Paper 554-C. 30 pp.

Hanks, H. G. 1883. Report on the borax deposits of California and Nevada. Calif. Min. Bur. Rept. 3, pt. 2. 111 pp.

Hardman, G., and M. R. Miller. 1934. The quality of waters of southeastern Nevada drainage basins and water resources. Univ. Nevada. Agr. Exptl. Sta. Bull. 136. 62 pp.

Hazzard, J. C. 1937. Paleozoic section in the Nopah and Resting Springs mountains, Inyo County, California. Calif. J. Mines and Geol., 33:273-339.

Herodotus. The Persian wars. Tr. by George Kawlinson. New York: Modern Library.

Hewett, D. F. 1955. Structural features of the Mojave Desert region (California). *In* A. Poldervaart, ed., Crust of the earth—a symposium. Geol. Soc. Am. Spec. Paper 62, pp. 377-390.

Hopkins, D. M., and others. 1955. Permafrost and ground water in Alaska. U.S. Geol. Survey Prof. Paper 264-F, pp. 113-146.

Hubbs, C. L. 1948. The Great Basin. Zoological evidence: correlation between fish distribution and hydrographic history in the desert basins of the western U.S. Bull. Univ. Utah, vol. 38, no. 20.

Hunt, Charles B. 1959. Dating mining camps with tin cans and bottles. GeoTimes. (Am. Geol. Inst.), 3(8).

————. 1974. Natural regions of the United States and Canada. San Francisco: W. H. Freeman and Company. 725 pp.

Jaeger, E. C. 1950. Desert wildflowers. Stanford: Stanford University Press. 322 pp.

————. 1957. A naturalist's Death Valley. Death Valley 49ers, Inc. Publ. no. 5. Palm Desert: Desert Magazine Press. 68 pp.

Jennings, C. W. 1958. Geologic map of California, Death Valley sheet. Olaf P. Jenkins, ed. Calif. Div. Mines. Scale 1:250,000.

Johnson, B. K. 1957. Geology of a part of the Manly Peak quadrangle, California. Univ. Calif. Dept. Geol. Sci., 30(5):353-423.

Kirk, Ruth. 1956. Exploring Death Valley. Stanford: Stanford University Press.

Kohler, M. A., and others. 1955. Evaporation from pans and lakes. U.S. Weather Bur. Research Paper 38. 21 pp.

Kroeber, A. L. 1925. Handbook of the Indians of California. Bur. Am. Ethnol. Bull. 78. Washington.

Lawson, A. C. 1915. The epigene profiles of the desert. Univ. Calif. Dept. Geol. Bull., 9:23-48.

Lee, Bourke. 1932. Death Valley men. New York: Macmillan. 319 pp.

Longwell, C. R. 1926. Structural studies in southern Nevada and western Arizona. Geol. Soc. Am. Bull., 37:551-583.

Lotze, Franz. 1938. Steinsalz und Kalisalze, Geologie. Berlin: Gebrüder Borntraeger. 936 pp.

Lucas, A. 1948. Ancient Egyptian materials and industries. 3d ed., London. 570 pp.

Mabey, D. R. 1960. Regional gravity survey of part of the Basin and Range Province. *In* Short papers in the geological sciences. U.S. Geol. Survey Prof. Paper 400-B, pp. B283-B285.

McAllister, J. F. 1952. Rocks and structure of the Quartz Spring area, northern Panamint Range, California. Calif. Div. Mines Spec. Rept. 42. 63 pp.

————. 1956. Geology of the Ubehebe Peak quadrangle, California. U.S. Geol. Survey Geol. Quad. Map CQ-95.

McDowell, S. D. 1974. Emplacement of the Little Chief stock, Panamint Range, California. Geol. Soc. Am. Bull., 85:1535-1546.

Malde, H. E. 1961. Patterned ground of possible solifluction origin at low altitude in the western Snake River Plain, Idaho. U.S. Geol. Survey Prof. Paper 424-B, pp. B170-B173.

Manly, W. L. 1894. Death Valley in '49. Los Angeles: Borden Publishing Co. (repr. 1949). 524 pp.

Maxson, J. H. 1950. Physiographic features of the Panamint Range, California. Geol. Soc. Am. Bull., 61:99-114.

Meinzer, O. E. 1922. Map of the Pleistocene lakes of the Basin-and-Range Province and its significance. Geol. Soc. Am. Bull., 33(3):541-552.

———. 1927. Plants as indicators of ground water. U.S. Geol. Survey Water-Supply Paper 577. 95 pp.

———. 1928. Outline of ground-water hydrology with definitions. U.S. Geol. Survey Water-Supply Paper 494. 71 pp.

Mendenhall, W. C. 1909. Some desert water places in southeastern California and southwestern Nevada. U.S. Geol. Survey Water-Supply Paper 224. 98 pp.

Merriam, C. H. 1898. Life zones and crop zones of the United States. U.S. Div. Biol. Survey, vol. 10. 79 pp.

Miller, R. R. 1948. The cyprinodont fishes of the Death Valley system of eastern California and southwestern Nevada. University of Michigan, Museum of Zoology Misc. Publ. 68. 155 pp.

Murphy, F. M. 1932. Geology of a part of the Panamint Range, California. Calif. Div. Mines. Mining in California, July-Oct., pp. 329-376.

Nelson, E. W. 1891. The Panamint and Saline Valley Indians. Am. Anthropologist, o.s., 4:371-372.

Noble, L. F. 1941. Structural features of the Virgin Spring area, Death Valley, California. Geol. Soc. Am. Bull., 52:941-1000.

Noble, L. F., and L. A. Wright. 1954. Geology of the central and southern Death Valley region, California. *In* R. H. Jahns, ed., Geology of southern California. Calif. Div. Mines Bull. 170, pp. 143-160.

Nolan, T. B. 1943. The Basin and Range Province in Utah, Nevada, and California. U.S. Geol. Survey Prof. Paper 197-D, pp. 141-196.

Pacific Coast Borax Company. 1951. The story of the Pacific Coast Borax Co. Los Angeles: Borax Consolidated, Ltd.

Palmer, T. S. 1952. Chronology of the Death Valley region, 1849-1949. Washington: Byron Adams. 25 pp.

Phalen, W. C. 1919. Salt resources of the United States. U.S. Geol. Survey Bull. 559. 284 pp.

Pistrang, M. A., and F. Kunkel. 1958. A brief geologic and hydrologic reconnaissance of the Furnace Creek Wash area, Death Valley National Monument, California. U.S. Geol. Survey Open-File Report. 63 pp.

Rankama, K., and Th. G. Sahama. 1950. Geochemistry. Chicago: University of Chicago Press. 912 pp.

Riley, C. V., and others. 1893. Report on insects. *In* Death Valley Expedition. U.S. Dept. Agr. Div. Ornithol. and Mammal., North America Fauna no. 7, pt. 2, pp. 235-268.

Robinson, T. W. 1952. Investigation of the water resources of the Nevares property in Death Valley National Monument, California. U.S. Geol. Survey Open-File Report. 21 pp.

_____. 1957. Determination of the flow of Saratoga Springs in Death Valley National Monument, California. U.S. Geol. Survey Open-File Report. 14 pp.

_____. 1958. Phreatophytes. U.S. Geol. Survey Water-Supply Paper 1423. 84 pp.

Rogers, M. J. 1939. Early lithic industries of the lower basin of the Colorado River and adjacent desert areas. San Diego Museum Papers no. 3. San Diego.

Shreve, F., and I. L. Wiggins. 1951. Vegetation and flora of the Sonoran Desert. Carnegie Inst. Washington Publ. 591, vol. 1.

Spears, J. R. 1892. Illustrated sketches of Death Valley and other borax deserts of the Pacific Coast. Chicago and New York: Rand McNally. 226 pp.

Stearns, R. E. C. 1893. Report on mollusks. *In* Death Valley Expedition. U.S. Dept. Agr. Div. Ornithol. and Mammal., North America Fauna no. 7, pt. 2, pp. 269-283.

Stejneger, L. 1893. Annotated list of the reptiles and batrachians collected by the Death Valley Expedition in 1891, with descriptions of new species. U.S. Dept. Agr., North America Fauna no. 7.

Steward, J. H. 1938. Basin-Plateau sociopolitical groups. Bur. Am. Ethnol. Bull. 120. Washington.

Ver Planck, W. E. 1956. History of borax production in the United States. Calif. J. Mines and Geol. 52:273-291.

_____. 1958. Salt in California. Calif. Div. Mines Bull. 175. 168 pp.

Walbridge, W. S. 1920. American bottles old and new. Toledo: Owens Bottle Co. 113 pp.

Wallace, W. J., and E. S. Taylor. 1955*a*. Early man in Death Valley. Archaeology, 8(2):88-92.

_____. 1955*b*. Archaeology of Wildrose Canyon, Death Valley National Monument. Am. Antiquity, 20(4):355-367.

_____. 1956. The surface archeology of Butte Valley. Contributions to California Archeology no. 1. Los Angeles: Archeological Research Associates.

Ward, F. N., and others. 1960. Geochemical investigation of molybdenum at Nevares Spring in Death Valley, California. *In* Short papers in the geological sciences. U.S. Geol. Survey Prof. Paper 400-B, pp. B454-B456.

Waring, G. A. 1920. Ground water in Pahrump, Mesquite, and Ivanpah valleys, Nevada and California. U.S. Geol. Survey Water-Supply Paper 450-C, pp. 51-81.

Waring, G. A., and E. Huguenin. 1917. Mines and mineral resources, Inyo County. *In* Report IV of State Mineralogist for biennial period 1915-1916, pp. 29-134.

Washburn, A. L. 1956. Classification of patterned ground and review of suggested origins. Geol. Soc. Am. Bull., 67(7):823-865.

_____. 1969. Weathering, frost action, and patterned ground in the Mesters Vig district, northeast Greenland. København: Meddelelser om Grønland. 301 pp.

Wasserberg, G. J., and others. 1959. Ages in the Precambrian terrane of Death Valley, California. J. Geol., 67(6):702-708.

Wauer, R. H. 1958. Check-list of birds, Death Valley National Monument, Inyo County, California. Death Valley: Death Valley Natural History Association.

Welles, R. E., and Florence B. Welles. 1962. The bighorn of Death Valley. Fauna of the National Parks of the United States, Fauna ser. no. 6. 242 pp.

Went, F. W., and M. Westergaard. 1949. Ecology of desert plants. Pt. III: Development of plants in the Death Valley National Monument, California. Ecology, 30:26-38.

Wheat, C. I. 1939*a*. The forty-niners in Death Valley: a tentative census. Los Angeles: Historical Society of Southern California. 16 pp.

_____. 1939*b*. Pioneer visitors to Death Valley after the '49ers. Calif. Hist. Soc. Quart., 18(3). 22 pp.

_____. 1939*c*. Trailing the forty-niners through Death Valley. Sierra Club Bull., 24(3). 37 pp.

Woollard, G. P. 1958. Areas of tectonic activity in the United States as indicated by earthquake epicenters. Am. Geophys. Union Trans., 39(6):1135-1150.

Wright, L. A. 1952. Geology of the Superior talc area, Death Valley, California. Calif. Div. Mines Spec. Rept. 21. 22 pp.

_____. 1954. Geology of the Alexander Hills area, Inyo and San Bernardino counties. Map Sheet no. 17. *In* R. H. Jahns, ed., Geology of southern California. Calif. Div. Mines Bull. 170.

_____. 1968. Talc deposits of the southern Death Valley—Kingston Range region, California. Calif. Div. Mines Geol. Spec. Rept. 95. 79 pp.

Wright, L. A., J. K. Otton, and B. W. Troxell. 1974. Turtleback surfaces of Death Valley viewed as phenomena of extensional tectonics. Geology (Geol. Soc. Am.), 2(2):53-54.

Yale, C. G. 1908*a*. Gold, silver, lead, and zinc; California; Oregon. U.S.
    Geol. Survey Min. Resources, 1907. Pt. I, pp. 187-234.
———. 1908*b*. Borax. *In* Mineral resources of the United States, calendar
    year 1907. Pt. II. Nonmetallic products. Washington, D.C.: U.S. Geol.
    Survey. Pp. 631-635.

Index

(Numbers in boldface italic refer to figures in text.)